How can we
SOLVE
Our SOCIAL
PROBLEMS?

How Can We
SOLVE
Our SOCIAL
PROBLEMS?

JAMES CRONE
Hanover College

PINE FORGE PRESS
An Imprint of Sage Publications, Inc.
Thousand Oaks • London • New Delhi

For information:

Pine Forge Press
A Sage Publications Company
2455 Teller Road
Thousand Oaks, California 91320
E-mail: order@sagepub.com

Sage Publications Ltd.
1 Oliver's Yard
55 City Road
London EC1Y 1SP
United Kingdom

Sage Publications India Pvt. Ltd.
B-42, Panchsheel Enclave
Post Box 4109
New Delhi 110 017 India

Printed in the United States of America

Library of Congress Cataloging-in-Publication Data

Crone, James A.
How can we solve our social problems? / James A. Crone.
    p. cm.
Includes bibliographical references and index.
ISBN 1-4129-4066-4 or 978-1-412940-66-5 (pbk.)
   1. Social problems. 2. Sociology. 3. Social problems—United States.
4. United States—Social conditions—21st century. I. Title.

HN18.3.C76 2007
320.60973—dc22   2006021824

This book is printed on acid-free paper.

06   07   08   09   10   11   9   8   7   6   5   4   3   2   1

Acquisitions Editor:	Benjamin Penner
Editorial Assistant:	Camille Herrera
Production Editor:	Diane S. Foster
Copy Editor:	D. J. Peck
Typesetter:	C&M Digitals (P) Ltd.
Proofreader:	Jenny Withers
Indexer:	Molly Hall
Cover Designer:	Michelle Kenny

Contents

Preface

How can we solve our social problems? For example, how can we solve the problem of poverty or the world's population problem? We have certainly made progress in dealing with certain social problems. We have food stamps for poor people. We have passed civil rights laws so that all Americans can go into public places such as restrooms, restaurants, hotels, and parks. We have health care for older people. However, can we do more? Yes, I believe we can do a lot more, and in the coming pages I discuss what more we can do.

I outline a number of realistic steps we can take to solve, or at least ameliorate or lessen, the severity of a number of social problems. Notice that I say "can" rather than "should." I present what we *can* do, but you will need to decide what we *should* do. As a sociologist, I cannot answer the "should" question for you because a science (natural or social) can deal only with the "is" part of phenomena such as description, causes, consequences, and prediction. Consequently, as you make your way through the book and read and think about how we can solve each social problem, I want you to be alert to what we can do and what you think we should do.

As you read the chapters, I remind you from time to time how challenging it will be to solve our social problems. In Chapter 2, for example, I discuss a number of barriers we face when we try to solve social problems. By discussing these barriers and making them conscious in your mind, I do not intend to make you feel that the situation is hopeless. What I do intend is to get you to have a realistic outlook. Neither a pessimistic and hopeless outlook nor a naive and innocent outlook will take us very far. Instead, we will need to have a realistic and pragmatic outlook, where our feet are firmly planted on the ground and where we are open to new ideas and possibilities that will give us hope that we can do something about our social problems.

About the Author

James A. Crone earned his Ph.D. in sociology at the University of Kansas. During the past 25 years, he has taught many sociology courses at Hanover College, where he is a member of the Department of Sociology and Anthropology. Over the years, he has served as chair of the department and as chair of a number of faculty committees on campus. He has been a finalist for the all-campus teaching award and was named Hanover College Faculty of the Year by the Greek Honor Society. He has published articles on sport and teaching sociology. He has served the larger community by being president of the Hanover Town Council and president of the local chapter of Habitat for Humanity, and he is currently serving on the Jefferson County Council. In his leisure time, he likes to play basketball with Hanover College colleagues and students, go hiking, and take walks. He has a son, Sasha, with whom he likes to go hiking, play tennis and basketball, and have fun taking trips together.

Acknowledgments

I thank all of the reviewers who made many constructive suggestions to improve this book: Chris Adamski-Mietus (Western Illinois University), Elizabeth M. Aranda (University of Miami), Tracy Citeroni (University of Mary Washington), Ilona Hansen (Winona State University), Constance L. Hardesty (Morehead State University), and Karen Tejada (State University of New York at Albany). I greatly appreciate their time, effort, and insight. At Pine Forge and Sage Publications, I also thank the acquisitions editor, Ben Penner, and his assistant, Camille Herrera, for their patience and continual support of me and my work, and I am grateful to Diane Foster and D. J. Peck for making the production process go smoothly. Finally, I thank the hundreds of students I have taught in social problems courses over the years who asked important and challenging questions and created enlightening discussions that, collectively, were the stimulus for my writing this book.

1

Preparing to Solve
Our Social Problems

Have you ever been concerned about a social problem? I imagine that you have. You may have been concerned about poverty, racial/ethnic inequality, or the inequality between men and women. Well, you are not alone. Many other Americans and people in other countries are also concerned about these and other social problems. One way to address these concerns is to read about and study social problems and think about how we might address these problems realistically. This book is a good place to start.

Before we begin to study specific social problems and consider how we can solve them, we need to learn a few things that will provide a base or foundation on which to build a more comprehensive understanding of social problems. With a fuller understanding, we will be more sophisticated in our study of social problems and better prepared to think about how we can solve these social problems. So, let us first build this foundation of understanding in Chapters 1 and 2, and then we will be ready to address our social problems.

What Is a Social Problem?

Before we turn to the main emphasis of this book, namely how we can solve our social problems, we need to define what a social problem is. One thing it is not is a personal problem that others do not experience. As C. Wright Mills, a respected American sociologist, pointed out, a personal problem can also be a social problem if a number of people experience the same personal

1

problem where certain social conditions are causing these people to experience the same personal problems. For example, many families experience poverty personally, but all of them are a part of a larger social pattern of unemployment, a social factor not caused by these families (Mills, 1959). Consequently, a key element in deciding whether something is a social problem is to discover how people's personal problems are related to the social conditions of a society.

Many social problems are at the societal level, such as poverty, racial/ethnic discrimination, and gender inequality. However, local communities can also define certain social conditions as social problems (Fuller & Myers, 1941). In addition to recognizing local and societal social problems, we are becoming more aware of global social problems such as the world's population problem, where many people do not have enough water to drink and enough fertile land to grow sufficient food. A social problem can therefore be at the local, societal, or global level.

Part of defining some social condition as a social problem is that we subjectively say to ourselves that something is wrong and that we believe it should be changed. For example, we say that we believe poverty is wrong and that we as a community, society, or world should do something about this. At the same time, Fuller and Myers (1941) asserted that social problems also need to have objective elements to them (p. 320). That is, we need to show that there is empirical evidence of a social problem. For example, when we collect data to show that poor people have lower incomes, lower quality of housing, and lower quality of life than do nonpoor people, we demonstrate that a social problem also has an objective element to it. We therefore need to have both subjective and objective elements in a definition of a social problem.[1]

Taking all of these things into consideration, we include the following as part of a definition of a social problem. First, certain social conditions cause personal problems. Second, social problems can be local, societal, or global. Third, social problems consist of both subjective perceptions and objective evidence. Hence, we use the following definition:

> A social problem exists when people subjectively perceive and have empirical evidence to show that social conditions combine at a local, societal, or global level to cause personal problems.[2]

History of Studying Social Problems

People have thought about social problems for a long time. In fact, this is one of the main reasons why the field of sociology began during the early

1800s. Early social thinkers (they were not called *sociologists* back then) during the late 1700s and early 1800s were concerned about all of the social changes that were occurring and wondered whether societies were falling apart. At that time, more and more people were moving to cities to get new kinds of jobs called *factory jobs*. Slums and crowded housing were created. Some people lost their jobs and experienced extreme poverty. The people who had no jobs and therefore no income would steal or rob at times, thereby making crime a social problem. As a result of these social changes, social problems of inadequate housing, poverty, and crime grew and became a typical part of the urban scene.

So much social change was occurring that some people such as Auguste Comte, a social thinker in France during the early 1800s, became conscious of and concerned about this social change and the resulting social problems.[3] From his viewpoint, society was falling apart due to too much disorder. Something needed to be done to bring some semblance of order and harmony to people's lives. He concluded that a new discipline was needed to study society—how it works, why it works that way, and where it is headed. He was concerned about what could be done about all of the social problems people were facing. Under these conditions, he created the new discipline of sociology to study society scientifically to see what could be done to make a more stable and orderly society in light of all the social changes. The new discipline of sociology was born out of Comte's concern for trying to understand order and change in society.

Later during the 1800s and early 1900s, like Comte, other social thinkers began to think about society in general and about social problems in particular. Emile Durkheim, a French sociologist, had concerns similar to those of Comte.[4] With the fall of monarchies and the apparent decline in the influence of religion, Durkheim wondered how modern society could keep any sense of order. German social thinkers, such as Karl Marx during the mid-1800s and Max Weber during the early 1900s, also became interested in how society worked and what social problems people faced.

Marx was greatly troubled by the increasing poverty and inequality he saw around him.[5] He was concerned about how people had factory jobs that were alienating because their jobs were so boring, they were paid so poorly that they could hardly survive, and yet they worked 12 hours per day, 6 days per week. Moreover, people did not have much choice. They either took alienating factory jobs or had no jobs and hence no means to sustain themselves. They were stuck in a social system that was brutal and inhumane, and they did not know what to do about their situation. As Marx pointed out, our ancestors created the very conditions that factory workers lived in—poverty, alienating jobs, and the lack of much choice in life. He further asserted that because our ancestors created these social conditions, we could—and should—change these social conditions. He helped us to realize, possibly more than any other social thinker, that we, as humans, created our social conditions and could therefore change our social conditions. We did not need to accept the existing social conditions as the only way to live.

Marx developed solutions to these problems that he thought would create a more humane society. He focused on what he thought was the main cause of many modern social problems: capitalism. He noted that although capitalism produced material benefits for many people and much profit for some people, at the same time it created alienating jobs, poverty, and much inequality. He concluded that we could do better than this and that we had the power to create a more humane and just society and world.

Another giant in the field of sociology was Max Weber.[6] He was concerned about the modern social problem of all the bureaucracies we live in and how these bureaucracies have considerable power over us. He predicted that individuals would feel helpless in the face of such large organizations. Consequently, Weber wondered how we would be able to solve the problem of our powerlessness in the face of these modern bureaucracies.

As you can see, from its beginning, the new discipline of sociology focused on the study of social problems and how these problems could be solved. Contemporary sociologists have the same focus. We too are curious about how society works, why it works the way it does, and what may happen in the future. We too are interested in how social conditions create social problems. And we too are interested in how we can change social conditions to solve our social problems.

Teaching About Social Problems in Today's World

Today in sociology, there are courses and textbooks devoted solely to the study of social problems. In these courses and the textbooks that are used for these courses, there is usually a focus on 10 to 15 social problems that are of current concern. Some of these problems are a concern and have been so for many years. For example, in our country we have been especially troubled over poverty, crime, and racial prejudice and discrimination. Other social problems have become of increasing concern to us within the past 50 years, including the growth in the world's population and the burden it plays on our limited resources, the deterioration of our global environment, and the continuing inequality between women and men in our country and throughout the world.

In our social problems courses, we focus on certain aspects of a social problem. For example, we describe a social problem such as how many people are being affected and where the problem is most prevalent (e.g., in cities, in the lower social classes, among women). We search for the causes of the problem, usually finding that there are a number of causes for each social problem but that some causes have larger impacts than others. We point out the intended consequences that are readily apparent but dig deeper to discover the unintended consequences that are not so apparent. We also make predictions as to what will most likely occur given certain social conditions. Finally, we discuss possible solutions, which is what this book addresses specifically.

How Can Sociologists Address Social Problems and Yet Remain Objective?

The study of social problems presents a dilemma for sociologists. The dilemma does not occur at the point of choosing a topic of research, of gathering facts in the research process, or of choosing certain methods of gathering data such as the survey, participant observation, and/or the interview. Sociologists, in general, agree that these kinds of activities are what sociologists need to do. These activities are an integral part of our being sociologists. The dilemma also does not occur at the point of discovering the causes or uncovering the unintended consequences related to each social problem. Sociologists, in general, seek to pinpoint causes and find unintended consequences.

The dilemma, however, does occur at the point of dealing with the solution part of social problems. That is, what should we say about the solving of social problems? Should we say what we personally think should be done? Should a group of sociologists come together to decide what should be done? Should we remain objective and not take a personal stance on what should be done and yet, in some way, contribute what we know about the solving of social problems?

Many sociologists contend that our role is to state only what is, that is, to study only what occurs, focusing on description, causes, consequences, and prediction but not saying what should occur.[7] They begin to feel uncomfortable when it comes to the solving of social problems because they worry that they or other sociologists may go beyond their role in being objective. They fear that if the general public no longer sees sociology as objective, sociology will lose its credibility.

The result would be that sociologists will be seen as just another interest group with its own vested interests instead of being seen as a group that the public and government can trust to report valid data and give objective knowledge on a subject so that others can make a more informed judgment as to what should be done. Consequently, a number of sociologists conclude that, rather than risk our credibility as an objective source of data, knowledge, and understanding, maybe it is better to stay away from the area of dealing with solutions. Instead, we need to leave this area to the policymakers of the society, such as members of Congress or state legislatures, and focus solely on descriptions, causes, consequences, and predictions.

There is another group of sociologists who see their role as not only stating what is but also stating what should be. Their belief is similar to that of Marx (1972), who said, "The philosophers have only interpreted the world, in various ways; the point, however, is to change it" (p. 107). That is, they ask what is the point of doing all of this studying of society, collecting mountains of data, and discovering causes and consequences if we do not take the next step to change the society for the better? They argue that if we study the problem more than anyone else in society and understand it the best, why not take the next step and say what should be done about it? After all, they say, we are the most expert on the study of social problems.

It seems to be a great waste of our knowledge and understanding of social problems if we cannot, in some way, venture into the realm of solving social problems. The key question then becomes the following: How can we study the solving of social problems and yet maintain our objectivity and credibility? Is there, in other words, a common ground to stand on for all sociologists?

Yes, there is a common ground on which we all can stand. On this common ground, there are at least five areas within which we can achieve the goal of contributing to the solving of social problems and yet remain objective and maintain our credibility.

One area related to the solving of social problems is to look at what sociologists know with regard to social patterns in social problems and how knowing these social patterns can help us to solve our social problems. We have a good idea of a number of social patterns that typically occur within various kinds of social problems.

Another area where sociologists can contribute to the solving of social problems and yet remain objective is to study the aspects of a social problem that prevent it from being solved. That is, sociologists can help us to become more aware of the barriers that prevent a social problem from being solved. Once we know these barriers, we can focus on how we can work around these barriers.

Another way sociologists can remain objective and yet contribute to solving our social problems is to study empirical examples of how social problems have been solved or lessened in the past or in other countries and reflect on how these solutions could be applied to the solving of other social problems. That is, what can we learn from the social problems that we have already solved, or at least ameliorated, that can be applied more generally to the solving of other social problems?

All sociologists also stand on a common ground when they make predictions about potential new social problems on the horizon and about where existing social problems are likely to go given current social policies and attempts to solve these social problems. By predicting new and emerging social problems and predicting where current social problems are likely to go in the near future, this information can be of considerable use to policymakers.

A final common ground for all sociologists is the ability to suggest various solutions and what their possible consequences might be for individuals, groups, communities, societies, and global social systems. Note that sociologists are not choosing a preferred solution. Instead, we are outlining what we think the possible solutions are, thereby helping policymakers to know more clearly what their options are. Contributing such knowledge could provide a great service to policymakers because this knowledge would give them more comprehensive information as to what they could do next.

What Social Problem Should We Solve First?

Given our limited resources, we cannot solve all of our social problems at one time. Consequently, it would help to have criteria to decide what social problem we should tackle first, then second, and so on.[8] This leads to the following question: "What are the most important criteria in deciding which social problem to address first?" The following criteria can help us to get started. One criterion is to assess the degree to which a social problem seriously endangers the lives of people; for example, one social problem causes little more than an inconvenience to people, whereas another social problem endangers their lives. A second criterion is the number of people being hurt by the social problem; for example, one social problem hurts hundreds of people, whereas another social problem hurts millions or billions of people. Using these two criteria, we can create the following two-by-two table as a visual means of deciding what social problem we should address first:

	Not Endanger Lives	*Endanger Lives*
Affect Some	Least serious	
Affect Many		Most serious

Applying these two criteria, we can conclude that a social problem that endangers the lives of people and affects many people (lower right cell) could be the one we address first, whereas a social problem that does not endanger people's lives and affects relatively few people could be the social problem we address last (upper left cell). As to the other two cells, I am not sure what would need to be chosen next. Maybe additional criteria will help us to answer this question. The point, however, is that given the limited resources we can apply to the solving of social problems, it is fruitful for us to use criteria to help us gain clarity as to what social problem we may want to tackle first and then second and so on.

How Might Sociological Theory Help Us to Solve Social Problems?

If we can understand the nature of a social problem better by applying sociological theory to it, we are in a better position to solve that problem. Let us first define what theory is before we begin using it. In sociology, theory has usually meant one of two things; either it is a collection of interrelated concepts and ideas, or it is a set of interrelated propositions that are applied to social phenomena to help us understand those social phenomena.[9] The "interrelated propositions" kind of theory needs a little bit of explanation, so we discuss this type of theory first before discussing a number of theories that consist of a collection of interrelated concepts and ideas.

With respect to a theory that is a set of interrelated propositions, you may ask, "What is a proposition?" A proposition tells us how one variable causes another variable to change. For example, say that you are interested in what causes poverty. You create a hypothesis, which is a predicted causal relationship between two variables, where you hypothesize that as the rate of unemployment (one variable) goes up, the rate of poverty (the second variable) goes up. Because unemployment is doing the "causing," we call this variable the *independent* variable. Poverty is the *dependent* variable because it is being influenced by unemployment.

We test our hypothesis to see whether what we think is occurring is really occurring. We collect data in some way, such as participant observation, the interview, and/or the survey, to see whether our hypothesis is supported by the data. If we test our hypothesis a number of times and find that, indeed, as the rate of unemployment goes up, the rate of poverty goes up, we begin to conclude that our hypothesis is probably true. As a note of caution, however, we might never be 100% sure because there could be another independent variable that is causing the rate of poverty to go up.

However, we can do what is called *controlling for other possible indepen-dent variables;* that is, if we have a random sample of data that does not allow for other variables to vary, such as a sample made up of all females (no males so that the variable of sex cannot vary), all African American women (no white, Hispanic, Native American, or any other kind of racial/ethnic group so that the variable of race/ethnicity cannot vary), all high school-educated women (no middle school- or college-educated women), only women with two children under 5 years of age (no women with no children, with one child, or with three or more children), all women who are 30 years of age (no women who are any other age), and so on, we now know that these variables cannot vary in our study.

We test our two variables. As these women in our sample increase in unemployment, do more and more of them also fall into poverty? If the data show that there is still a relationship (or correlation) between the rate of unemployment and the rate of poverty, we are closer to being confident in saying that unemployment is probably a cause of poverty. Once various sociologists in various studies find the same data in different studies, we begin to be cautiously confident that the independent variable in our hypothesis is a cause of the dependent variable in our hypothesis.

Once our hypothesis is firmly established, we can begin to say that it may be a theoretical proposition, that is, a hypothesis that not only is found to be true but also is more abstract; that is, it can apply to various social con-ditions in this society and other societies, and it can apply over time or to various times in history. In other words, the more a proposition applies to more diverse social conditions and the more it applies throughout more of history, the more abstract the proposition is. Ideally, we in sociology, as well as those in other social sciences, would like our theoretical proposi-tions to be more abstract so that they apply to more social conditions over more time. In reality, some theoretical propositions are very abstract and some are more concrete in that they apply to a limited number of social conditions over a smaller amount of time.

Let us go one step further and connect one theoretical proposition with another theoretical proposition, that is, where one of the variables is pre-sent in both propositions. When we begin to do this, we begin to call this *theory.* For example, let us relate the proposition of the higher the rate of unemployment, the higher the rate of poverty, to another proposition where either the rate of unemployment variable or the rate of poverty vari-able is one of the variables in the second proposition.

Let us say that our country is losing factories to other countries and that nothing else is taking their place. Factories are moving to Mexico, South Korea, Thailand, and China to get the benefit of paying lower wages and not needing to provide workers with any health and retirement benefits, thereby making more profit for the companies and making more money for those who own stock in those companies. This process has been occurring since the 1960s and is known as *deindustrialization.*

We can now make a second proposition that connects to our original proposition: The greater the rate of deindustrialization in a community or society, the greater the rate of unemployment in that community or society. Notice that we connected this second proposition with the original proposition by using one of the variables that occurs in both propositions, that is, the rate of unemployment. Here are the two propositions so that you can see how they are connected:

> *Proposition 1:* The greater the rate of deindustrialization in a community or society, the greater the rate of unemployment.
>
> *Proposition 2:* The greater the rate of unemployment, the greater the rate of poverty.

Now we could create more propositions that help us to further develop a theory, say of poverty. We could, for example, state the following proposition: The greater the rate of poverty, the greater the rate of homelessness. Notice that the rate of poverty becomes the independent variable, which we predict causes the rate of homelessness to increase.

Now that you know what a theory of interrelated propositions is, I want to share with you what I call a *theory of conflict and social change* as a way to help us better understand and solve our social problems.

A Theory of Conflict and Social Change: A Way to Better Understand and Therefore Solve Our Social Problems

In this theory, I have created a number of interrelated propositions that will help us to understand how social problems can be solved or at least ameliorated. In this theory, I have not included every variable that influences a social problem, but I have included a number of variables that I believe play a large role in what happens to social problems.

Before we talk about specific variables, I want you to realize that we, as humans, have socially constructed our reality.[10] That is, whether you realize it or not, we have created all kinds of social phenomena such as slave systems, various kinds of prejudice and discrimination, democracies and dictatorships, economic systems such as capitalism and socialism, and norms, values, and beliefs that are important in one culture or at one point in history and not present in another culture or at another time in history. We have created all of these things and yet, as Marx (1964) pointed out, these can come to control us—many times for hundreds or thousands of years. Over the years, we come to forget that someone back in history socially constructed these social phenomena and typically created ideologies, laws, and customs to legitimize these social creations. Over time, we come to accept society the way it is and live a certain socially constructed way of life for generations. It is as though we say to ourselves, "This is the way it is and has been, and this is the way it will always be."

Then something happens in society (I elaborate soon) to cause people to realize that what seemed to be beyond us as humans was actually constructed by humans and therefore could be changed. Once this realization occurs, there is the possibility that people can seek social change.

Given these initial ideas on the human constructedness of our social world, we can now proceed to create a theory of conflict and social change that consists of a number of interrelated propositions where each proposition consists of one variable influencing another variable. Taken together, these propositions create a theory of conflict and social change that will help us to understand the nature of social problems and solve these problems.

One key variable that seems to be related to many social problems is the amount of inequality in a community, society, or global social system. By *inequality*, I mean that some people, groups, organizations, and societies have more money (income and/or wealth), power, and prestige than do others.[11] There are many causes of inequality in contemporary society, but I assume that the following social factors explain much of the inequality that we have had, and currently have, in our society: various kinds of prejudice and discrimination, a capitalistic economy, and unequal exchange relationships that are supported by accepted norms, vested interests, ideologies, laws, and customs.

As you may already know, those who experience prejudice and discrimination will have fewer chances and opportunities, hence creating more inequality. In capitalism, some people are owners and make a profit, whereas others are workers and make a wage. Typically, owners make a lot more money in making profit than workers do in earning wages, thereby creating inequality. In every society, there are exchanges of goods and services. Sometimes those exchanges are unequal and the norms that are created say, "This is the way it should be; when we exchange something for something, you get this and I get that." This kind of norm is called a *norm of reciprocity*, where people accept the prevailing exchange relationships as legitimate

(Gouldner, 1960). Once people accumulate more money, power, and prestige, they develop vested interests in keeping, or even increasing, these resources. Given these vested interests, many times they will socially construct ideologies, laws, customs, and informal norms to justify and maintain the resources that they have accumulated. If these ideologies, laws, customs, and informal norms are accepted as legitimate by the rest of the people, those who have the extra resources will retain these resources. All of these socially constructed social factors together create a certain amount of inequality in a given society. Consequently, here are the first theoretical propositions dealing with what contributes to inequality in a society:

1. The more prejudice and discrimination that occur, the more inequality will occur.

2. As capitalism develops, some people will make profit while others will earn wages, thereby creating more inequality.

3. As unequal exchanges occur, more inequality will occur.

4. As norms of reciprocity that justify unequal exchanges occur, more inequality will occur.

5. Those people who make and accumulate more money, power, and prestige will be more likely to develop vested interests in maintaining, or even increasing, the existing amount of inequality, thereby leading to more inequality.

6. Those people who make and accumulate more money, power, and prestige will be more likely to create ideologies, laws, customs, and norms to maintain, or even increase, the existing amount of inequality, thereby leading to more inequality.

Another key variable is the amount of opportunity people have. Some people have many opportunities, whereas others have few opportunities. For example, in the cases of poor people, African Americans, and women, we know that these people do not start at the same "starting line" of equal opportunity as do others such as middle-class Americans, white Americans, and male Americans. Consequently, here is the next theoretical proposition:

7. The more inequality there is in a social system (group, organization, community, society, or global social system), the less opportunity people who are at the bottom of the system of inequality will have.

As people have less opportunity, they will be less likely to be upwardly mobile. This situation is especially important in a class-type society versus a caste-type society. Whereas people are socialized to "stay in their place" in a caste or caste-like society, people are socialized to "get ahead" in a class-type society. In a class-type society such as the United States, when people have less opportunity to get ahead, they will be less upwardly mobile than people who have more opportunity. In other words, given the structural situation of less opportunity, we should predict that people in this situation will experience less upward mobility. Consequently, we add another proposition:

8. The less opportunity people have, the less upwardly mobile they will be compared with people who start life with more opportunity.

Assuming that nothing else influences people at the lower end of a system of inequality, people are likely to accept the prevailing system of inequality as legitimate and will therefore live within this system and do nothing to attempt to change it. However, if these people are able and begin to communicate with each other about their respective conditions, and if a charismatic leader communicates dissatisfaction regarding the current system of inequality, people can begin to develop an awareness or consciousness of their unequal situation. These additional variables of the ability to communicate and the presence of a charismatic leader suggest two more propositions:

9. The more people at the bottom of the system of inequality can communicate with each other about their situation, the more aware they will become of their situation.

10. The more a charismatic leader communicates the unequal situation that people experience at the bottom of a system of inequality, the more aware they will become of their situation.

As people are able to communicate with others who are in the same unequal situation as they are, and as a charismatic leader is successful at communicating this unequal situation, the situation is ripe for people to begin to develop a feeling of unfairness. Hence, there is a need for the next proposition:

11. The more aware people become of their unequal situation, the more likely they are to develop a feeling of unfairness.

As people continue to talk about their unequal situation and develop a feeling that their situation is unfair, they are more likely to do something that those in power and those who want to maintain the status quo do not want them to do, that is, to question the legitimacy of the existing social construction of reality. In other words, people at the bottom of a system of inequality may come to question ideologies, laws, customs, and informal norms and to question why other people have so much money, power, and prestige. They may say to themselves, "Why do we need to live like this? Why should we have so little while they have so much?" This kind of thinking can lead people to question the very foundation of a society.[12] With this in mind, let us add another proposition:

12. The more people develop a feeling of unfairness about the current system of inequality, the more likely they will begin to question the legitimacy of a social system and possibly various aspects of the society in general, that is, how that society socially constructs its reality.

Given these conditions, we may begin to see the formation of a group or an organization that begins to organize itself around solving a social problem. For example, in 1955, Martin Luther King, Jr., a soon-to-become charismatic leader, and members of the African American community of Montgomery, Alabama, began to focus their efforts on forming an organization to boycott the segregated bus system. Hence, we need another proposition at this point:

13. The more people question the legitimacy of a social system, the more likely they will form a group or an organization to take action to address a social problem.

Once people have formed a group or an organization, they are more likely to take some kind of action to change the current social situation, meaning that they will have some kind of conflict with people, groups, and organizations that do not want to change the social system. The conflict can be one of two types: peaceful or violent. In reality, there can be both types—on the part of people who want change and on the part of people who do not want change. People seeking social change and the solution or amelioration of their problem can use peaceful means of conflict such as writing letters to legislators, attending town meetings to express their grievances, walking in protest marches, and boycotting stores, buses, and restaurants. They can also use more violent methods such as rioting. Those who want to keep the status quo can urge the seekers of change to "go by the law" and "stay in your place" and use laws to enforce the status quo. If need be, they can use governmental force in the form of the police or military or use informal methods of threat, torture, and murder such as the Ku Klux Klan and other racist organizations did to maintain the status quo.

With conflict (peaceful and/or violent) comes the possibility that something can change in the social system and possibly solve or ameliorate a
social problem. Once conflict has occurred, we may notice various kinds of
social change. For example, we might see social change in ideologies, for
example, from a segregation ideology to an integration ideology. We might
see laws change, for example, the Civil Rights Act of 1964 where public
places were open to any American and the Title IX Act where females were
to be given equal opportunity in schools receiving funds from the federal
government. We might see change in informal norms where people are
expected to be friendlier toward each other and more respectful of each
other. We might see a change in the redistribution of money, power, or
prestige or a combination of these three dimensions of inequality in the
form of new social services offered, for example, the creation of social security, Medicare, and Section 8 subsidized housing. We might see more
opportunities made available to people so that they can be more upwardly
mobile, for example, programs such as Head Start and grants and loans for
poorer people to get to go to college or trade school. Hence, we need two
final propositions:

14. The more a group or an organization carries out conflict (peaceful and/or
violent) with the intent of solving or ameliorating a social problem, the more
likely one or more kinds of social change will occur and result in a new
social construction of reality.

15. To the degree that social change occurs with the intent of solving or ameliorating a social problem, the more likely a social problem will be solved or
ameliorated.

Causal Model: A Picture of Our Theory

Now let us create a picture of our theory to visualize how the propositions
are tied together. When we create a picture of a theory, this is called a *causal
model* (Turner, 1991, pp. 15–28). That is, we show how the preceding
propositions are related by connecting them with arrows indicating the
direction of causality (Figure 1.1).

As you can see from analyzing Figure 1.1, the variables at the left of the
causal model influence the variables to the right. As you can also see, there
will be some kind of conflict that can result in some type of social change.
Part of this social change can involve the solving or ameliorating of a social
problem. Given that these theoretical propositions are valid,[13] we should
be better able to analyze social problems and how we could solve these

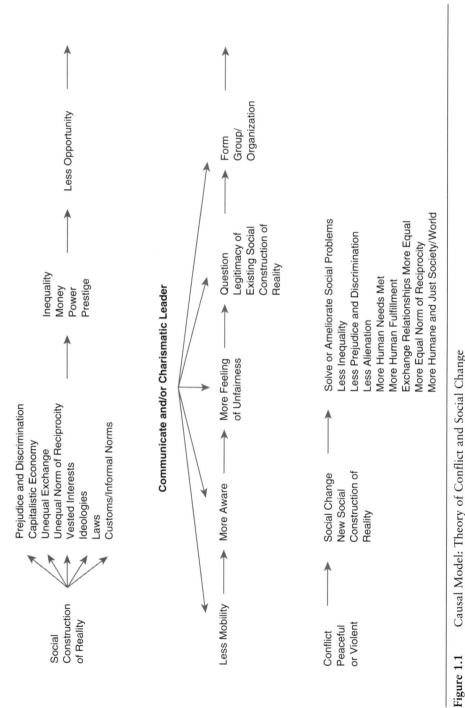

Figure 1.1 Causal Model: Theory of Conflict and Social Change

problems because we will be more conscious of the key variables that are a part of social problems. The more conscious we are of these key variables, the more we can make changes in these variables as a way to solve our social problems.

My Own Theoretical Orientation

Professors in sociology, like professors in other social and natural science disciplines, tend to favor one theoretical perspective over others with respect to analyzing certain social phenomena because they think that a particular theory will provide a fuller explanation. When it comes to the analysis of social problems and how these problems can be solved, I favor the conflict theoretical perspective that I just outlined because I think the reason why we have so many social problems has a lot to do with the interrelations among money, power, inequality, vested interests, ideologies, and legitimation—all of which are integral parts of a conflict theoretical approach. So, as you make your way through this book, you will notice that I continually use conflict theory, especially as formulated in the theory, propositions, and causal model that we just discussed. I also use other sociological theories, from time to time, if I think that they will help us to increase our understanding of social problems. With this in mind, I want to share with you these other sociological theories.

Other Sociological Theories to Help Us Understand and Solve Social Problems

Functional Theory

One theory that is well known in sociology is functional theory, also known as structural–functional theory. Although it was originally developed by Parsons (1951), I use just part of this theory as elaborated by Merton (1967) in his work, *On Theoretical Sociology*. In this work, Merton created the concepts of manifest and latent functions and dysfunctions. Here I define, explain, and combine these concepts and discuss how they can help us to understand and solve social problems.

I define *function* as consequences that increase the survival of a social system (a group, an organization, a society, or a global social system). For example, let us say that Company A gets a major contract with Company B to produce a certain product for Company B. This is functional for

Company A because getting the contract has the consequence of increasing its survival. Notice that I am talking about increasing the survival of Company A and that I did not make the value judgment that this was positive or negative or right or wrong.

In other words, function is not dealing with positive or negative or with right or wrong; rather, it is dealing with whether or not something has the consequence of increasing a social system's survival. Let me give an example in American history to show the distinction. We can hypothesize that the laws in the southern states during the 1700s and early 1800s had the consequence of increasing the survival of the slave system. These laws were therefore functional for the survival of the slave system. However, whether the slave system was positive or negative, or whether it was right or wrong, is another question. So, we might personally believe that the slave system was negative and wrong. However, as sociologists, we will want to discover how something, whether we like it or not, is functional for a certain social system, that is, how something increases the survival of that social system.

Another concept, *dysfunction*, deals with consequences that decrease the survival of a social system. For example, prejudice and discrimination against a group of people will have the consequence of decreasing the survival of that group. Given prejudice and discrimination, this social group will have fewer choices of jobs, fewer chances to get more education so as to get better paying jobs, and fewer chances to join various kinds of voluntary organizations such as churches, country clubs, and other voluntary organizations that provide networks of getting to know people who have connections that help people to get jobs, give advice on how to get ahead, and give financial support so that they can get ahead. In other words, prejudice and discrimination will have consequences for decreasing the survival of a certain group of people by decreasing their chances at gaining more money, power, and prestige.

A function can be either *manifest* (i.e., intended) or *latent* (i.e., unintended). For example, when there is intended prejudice and discrimination against Group A by Group B, the members of Group B will increase their survival because the intended prejudice and discrimination will typically have the consequence of giving them more choices, chances, and connections and therefore increase their survival. Hence, a manifest function states that there are intended consequences that increase the survival of a certain social system. Again, notice that I am not saying that this is positive or negative or that it is right or wrong; rather, I am saying that something has the consequence of intentionally increasing the survival of some social system.

A manifest dysfunction, on the other hand, states that there are consequences that are intended to decrease the survival of a social system. For example, using the preceding example again, whereas Group B's intended prejudice and discrimination have consequences of increasing its members' survival, it has consequences of decreasing the survival of Group A members who are being prejudiced and discriminated against. Thus, what is functional for one social system can be dysfunctional for another social system.

Finally, something can be a latent function or a latent dysfunction for a certain social system, that is, a function or a dysfunction can have an unintended consequence for a certain social system. For example, many African Americans migrated from the South to northern cities during the 20th century in search of better paying jobs and more freedom. That is what they intended to do. However, what they did not intend to do, but what ultimately happened, was that as they congregated in certain inner-city neighborhoods in northern cities because of the intended racial segregation by whites, blacks became a political power by being able to vote in black mayors, city council members, and members of Congress. Increasing their political power was not what they originally intended when they moved north. They moved north because they intended to get better paying jobs and experience relatively more freedom. However, their migration north also turned out to be a latent function, that is, an unintended consequence that increased their survival in that they began to have more political power in local, state, and national governments.

So, in dealing with many social problems, such as racial and gender inequality, poverty, the world's population problem, and the world's environmental problem, you can see how it will benefit us to use the ideas of manifest and latent function and dysfunction as a way to help us understand our social problems and how to solve them.

Symbolic Interaction Theory

Another well-known theory in sociology is symbolic interaction theory. Whereas functional theory deals with how there are functions and dysfunctions on social systems, symbolic interaction theory focuses on the micro level of society on interactions between and among small numbers of people. We use ideas such as Cooley's (1902) concept of the "looking glass self," where he pointed out that we will many times look to others to see how they see us and judge us as a way to decide how we see and judge ourselves. How have African Americans, women, and poor people at various times in American history been seen and judged? And how did they come to see and judge themselves?

These looking glass selves can help us to understand social problems in that the perception of the self can be a barrier to people being able to get ahead. That is, not only do objective conditions of prejudice, discrimination, and poverty affect people's chances of getting ahead in life, but also the kinds of selves people are socialized to have can act to hold them back. Hence, we will need to apply theoretical ideas from symbolic interaction theory to help us gain a greater understanding of our social problems and what steps we can take to solve these problems.

Anomie Theory

Another well-known sociological theory I use that will help us to understand social phenomena such as crime and deviance is anomie theory by Merton (1938). In his article titled "Social Structure and Anomie," Merton noted that most people are what he called *conformists* in that they have the legal opportunities available to them to gain the goals of society such as money, power, and prestige. However, there are others in society, due to their social conditions, who have blocked opportunities as they attempt to attain the goals of society. Due to these blocked opportunities, they are more likely to use illegal means to attain the goals of society. Merton called these people *innovators*. These people become innovators because the social conditions in which they live make it much more difficult for them to use legal means to achieve the goals of society. Merton's theory will help us to understand the deviance that occurs in a society due to the blocked opportunities that the existing social structure creates for certain groups of people. Using his theory will help us to reflect on how we can change the social structure as a way to give more legal opportunities to people. As we give more legal opportunities to people who do not currently have these opportunities, we will unlock another door that will help us to solve our social problems.

Reference Group Theory

Another theory that will help us to understand our social problems and how to solve them is reference group theory (Merton, 1968a, 1968b). This theory notes that we all refer to certain groups in our lives to decide how to think, feel, and act. There are usually many different groups we use as reference groups such as our family, a group of friends, a team, a fraternity or sorority, or some other social organization. While we are members of some of these groups, we are not members of others—yet we can still refer to them to decide how to act, think, and feel. These groups, whether or not we are members of them, can also be positive or negative in our eyes. Regardless, we still refer to them. For example, we may refer to our family and how we were socialized to decide how to act in a certain way; for example, you were taught to be considerate of all people regardless of their race, ethnicity, religion, or sexual orientation, and you proceed to act in that way. Others were taught by their families to be prejudiced and to discriminate, and—not surprisingly—they think, feel, and act differently from how you do.

Reference group theory does not say that you act exactly the way you were taught by the reference groups you have, but it does say that you will refer to these reference groups in deciding how you are going to act, think, and feel. Remember that outside social factors in our lives do not totally determine how we think, act, and feel, but they do tend to have a substantial influence on us even though we are not always aware of such influence. This is what Durkheim (1938) meant when he said that society is external to us and yet coercive on us. That is, society's laws, values, beliefs, and customs are outside of our minds when we are born, but through the process of socialization, these social phenomena are put into our minds and influence us more than we realize. Most of the time, we go through our daily lives and do not realize how much we are influenced by what we have been taught by the reference groups we have.

Differential Association Theory

Another sociological theory that I use from time to time to help us gain greater clarity, especially with respect to social problems such as crime, drugs, and the problems of families, is Sutherland's (1940) differential association theory. Sutherland, in focusing on crime, hypothesized that criminality "is learned in direct or indirect association with those who already practice the behavior" (p. 10) and that "those who learn this criminal behavior are segregated from frequent and intimate contacts with law-abiding behavior" (pp. 10–11). Knowledge of this theory will help us, for example, to understand why people deviate from the norms of the society and commit crimes and other acts of deviance (e.g., a member of the Ku Klux Klan being more willing to discriminate because he or she learns this behavior in association with others who have learned this behavior). We apply this theory to various social problems to help us better understand how the associations people have can place them in the social problems they experience and that a change in their associations may be part of the answer to their no longer experiencing these social problems.

Exchange Theory

Finally, one more sociological theory that I use from time to time is exchange theory by Blau (1964).[14] Blau asserted that we not only exchange material goods but also exchange nonmaterial things such as love, time, attention, support, and sympathy. When you think about it, many of our interactions with other people, groups, and organizations are interactions that include some kind of exchange. Besides going to a car dealer and exchanging money for a new car or going to a food store and exchanging money for groceries, we also give our time to someone or some group and typically receive something in return such as approval, acceptance, love, or attention.

A parent gives us love, and we give back love, time, or trust to that parent. We give a girlfriend or boyfriend time, attention, and love and hope that he or she gives us these things in return. In fact, a relationship tends to continue based on mutual giving and receiving. However, the relationship is in jeopardy if, for example, you give love, time, and attention to your boyfriend or girlfriend, but he or she no longer gives these things to you. When someone gives something to someone else (e.g., love, time, attention) and expects something similar in return but does not receive anything in return, this is called *breaking the norm of reciprocity* (Gouldner, 1960). That is, whether we realize it or not, we create expectations of exchange when we create various kinds of relationships—whether it be parent–child, boyfriend–girlfriend, husband–wife, teacher–student, coach–player, employer–employee, or another relationship. In other words, we create norms in exchanges where if we give something, we expect something in return. The exchange might not necessarily be equivalent, but there is an expectation of receiving something in return. For example, if you are a college student and say hello to someone you know as you walk across campus, you probably expect a hello in return. If that person looks the other way and ignores you, you will probably become immediately conscious that there was a breaking of a norm of reciprocity between you and the other person.

Bob likes Sue. Sue likes Bob. They want to give each other time, attention, sympathy, and love. Bob, however, is also a football player and spends 2 to 3 hours each day practicing or playing in a game. He is also expected to spend time with his fraternity brothers. He has little time left to be with Sue. Sue, on the other hand, has lots of time to be with Bob. She is not in a sorority or on a team. She wants to spend more time with Bob. Bob cannot spend as much time with Sue as she wants. He cannot be with her on Friday nights, on Saturday afternoons, or at other times. Sue pays a lot of attention to Bob (calling him and asking him, "When can we be together again?"), but Bob can spend only a certain amount of time with Sue. Sue's feelings are hurt because she tries to give so much time and attention to Bob, but he does not reciprocate with the same amount of time and attention. Sue can feel that the norm of reciprocity is broken, and this can place the relationship in jeopardy.

Although the example I just used is a romantic example, exchange theory and the norm of reciprocity can also be applied to social problems. Millions of people in our country work 40 hours per week but get paid wages considerably below the poverty line. The people involved may conclude that this relationship is an unfair exchange and may go on strike or take some other social action because they believe that the norm of reciprocity has been broken. Hence, the breaking of the norm of reciprocity can lead to conflict and some kind of social change. So, exchange, the norm of reciprocity, and the breaking of the norm of reciprocity all can be integral parts of a social problem and the possible solving of that problem. Consequently, you can see how exchange theory can be a part of my theory of conflict and social change outlined earlier where various inequalities of exchange or the breaking of expected exchange relationships can be causes for conflict and subsequent social change.

Concluding Thoughts on Theory and Social Problems

Note that as we work our way through this book, we continually use the theory of conflict and social change as a main means of understanding and solving our social problems. But we also apply the other theories just outlined to give us additional insight into a particular social problem. As you make your way through this book, you will see how beneficial it will be to apply theory to social problems so as to better understand and solve these problems.

Addressing Social Problems Within Capitalism

I want to make a comment on the context within which we address our social problems. I suggest what we can do to solve or ameliorate our social problems within the context of a U.S. and worldwide capitalistic economy. The reason why I say this is that I think that Americans, as well as people throughout the world, will live our lives, at least for the immediate future, within a capitalistic economy. How long will this be? I do not know, and I imagine that no one else does either.[15] If we look at capitalism during the past 300 to 400 years, it has continued to evolve, adapt, and adjust to the times. Will it continue to do so? I do not know. However, given its ability to adjust so far, its current pervasive influence in the world, and the legitimacy it carries among many powerful people, I predict that we will continue to live within such an economy for some time to come. So, with this in mind, my book is about what we can realistically do to solve our social problems within the context of a U.S. and worldwide capitalistic economy.

The Next Chapter

Before we discuss our first social problem, I want you to focus on the next chapter to become aware of three things: (a) the barriers to solving our social problems, (b) the possibilities for solving our social problems, and (c) how sociology can help us to solve our social problems. So, read and reflect on the next chapter as a way to prepare yourself to tackle the social problems we address later.

Questions for Discussion

1. Should we, in sociology, go beyond our role of being objective and say what we think should happen?

2. How much should sociologists be a part of changing society?

3. What might be other criteria, besides the number of people being affected and the potential for endangering lives, that we could use to decide which social problem we should tackle first?

4. Can we do without bureaucracies such as the government and large corporations?

5. If we need to have bureaucracies, are there some ways that we, as individuals, cannot be so powerless against them and be able to feel that they are not so impersonal to our human needs?

6. Why do people, at times, accept their unequal situations as long as they do?

7. What are some ways that we can become aware of social problems in our current society and world?

8. Why is there not much social change even when a number of people want it?

9. How can our theory apply to a social problem that we have today?

10. What are some things we can do in society and the world that can help us to have peaceful conflict versus violent conflict?

2

Barriers, Possibilities, and How Sociology Can Help

There are a number of barriers that prevent us from solving our social problems or that at least slow us down in making progress toward solving our social problems. Yet at the same time, there are possibilities for solving our social problems. Let us point out these barriers and possibilities so that you will become more conscious of a number of the key factors that influence the solving of social problems. Note that our discussion of these barriers and possibilities might not, and probably does not, exhaust all of the factors that influence the solving of social problems. However, these factors are probably what many sociologists would agree are key factors. Finally, I want to share with you how sociology can contribute to the solving of our social problems.

Barriers to Solving Our Social Problems

Power

One of the largest barriers to the solving of our social problems is the existing power structure, power elite, and powerful corporations that together will not want the solving of social problems if this means that people in positions of power and powerful organizations—especially large multinational corporations—will lose their power. Put simply, once people and organizations have power, they do not want to give it up along with all of the material and nonmaterial amenities that go along with having power.

Throughout history and in today's world, people who have power do not want to give it up. They do all kinds of things such as constructing ideologies (e.g., religious, economic, political) to justify their staying in power, giving some of their money and resources in the form of wages and social services as a way to stay in power, creating laws to legitimize the existing social conditions and hence the existing conditions of inequality, and using force (if needed) in the form of the police and military to maintain the status quo.

People and organizations that have power many times get their power from the money they have. So, whoever makes a lot of money, especially in the capitalistic societies of today's world, can and often will wield much more power than the rest of us. Of all the barriers that we discuss, this barrier of power and who has it will constantly play a major role in how we deal with our social problems. In other words, to understand why we do not readily solve or diminish the severity of our social problems, we need to discover and continually be aware of who has power and how they use their power.

What I am saying is not anything new. Many people throughout history have been aware of the relationship between power and what occurs in society. For example, during the mid-1800s, Marx and Engels (1978) wrote in *The Communist Manifesto,* "The ruling ideas of each age have ever been the ideas of its ruling class" (p. 40). Whatever the type of society, whether it be ruled by kings or dictators or owners of capital in a modern capitalistic society, those who have the power will tend to go after their own vested interests, which will usually be to remain in power and retain a disproportionate share of the material comforts of life, and hence will not want to solve social problems if that means they will lose their power and their way of life. They may support the alleviation of deplorable conditions of the masses so long as they do not lose their power and privileges (Lenski, 1984).

During the middle of the 20th century, Mills (1956) hypothesized that there was a power elite in American society that made a disproportionate amount of the decisions in our society. Mills hypothesized that people in the top positions in the corporate world, the military world, and the political world are the "movers and shakers" of our society. Kerbo (2006) asserted that sociologists need to speak of a new social class, namely the corporate class. He noted that the very largest corporations in our country have huge amounts of power such as the ability to lay off workers in one community, move a plant out of the country, decide how taxpayers' money is spent, and influence the making of laws in communities, states, and nations. To show the potential power that the largest 100 American corporations can yield, he noted that there "are about 200,000 industrial corporations operating in our country today, but almost 75 percent of all the industrial corporate assets are in the hands of 100 (0.0005 percent) of these corporations" (p. 179). In other words, there is a huge concentration of capital in relatively few hands in the United States. Moreover, Kerbo went on to report that of the top 5 stock-voting positions in the largest 122 corporations in the United States (5 × 122 = 610 positions in all), "just 21 investors accounted for more than half of all these top five stock-voting positions" (p. 184). So, a few people and a few corporations wield vast amounts of power as a result of their control over huge amounts of industrial assets. Today, in the 21st century, this is where much power appears to be shifting—to the very largest multinational corporations.

The bottom line for the solving of our social problems is that the vested interests of these very large and powerful corporations that seek survival and profit first and foremost might not, and many times do not, coincide with the solving of various social problems of our country. No doubt, this barrier will continue to be a daunting challenge as we seek to solve our social problems.

Vested Interests

Another big barrier to solving social problems is the idea of vested interests. Vested interests means that people or organizations will support, defend, and preserve what is to their benefit and will struggle against what can hurt them—their income, wealth, power, prestige, and way of life. They may even be sympathetic to changes that could be made that help others. But if the proposed social policies and subsequent changes go against their own vested interests, they still may be against any social policy that could solve or decrease social problems. As Blumer (1971) suggested some years ago, "A social problem is always a focal point for the operation of divergent and conflicting interests" (p. 301).

Here is an example that occurred throughout the 20th century. What would you do if you were in this situation? It is just after World War II when Will, who is white and has a family, has been back from the war for 2 years. While serving in Europe during the war, Will met Al, an African American soldier. They found out that they both were from Atlanta, Georgia, and over time they became good friends during the years they were in the war.

Soon after the war, Will and his family bought a house in a middle-class, all-white neighborhood in Atlanta. Al's family visited Will and his family and really liked the neighborhood. Al decided that he would try to buy a house that had a "for sale" sign in the front yard just a few houses away from Will's house. Al talked with the realtors (i.e., people who sell houses for the home owners). Al was ready and financially able to buy the house.

However, the realtors felt social pressure on them not to sell to an African American family because it was thought that an African American family moving into an all-white neighborhood would bring down the value of all the houses in the neighborhood given that there would be a fear that other African American families would want to move into the neighborhood and that white families would panic and sell their houses below market prices. Also, besides the fear of losing the value of their homes, some white families were still prejudiced and did not want to live near African American families or want their children to go to the same school as African American children.

Notice that from a friend perspective, Will would have no reservations about Al and his family moving into the neighborhood. However, from Will's vested interests of wanting to keep up the value of his house, he may be ambivalent and somewhat fearful of Al moving into the neighborhood.

This story that I made up occurred numerous times in our country over the past 100 years. Even when people are personally sympathetic, their own vested interests may still outweigh their sympathies. The same goes for getting a job, attending a school, belonging to a country club, or even being a member of a church. Sympathies can be strong, but vested interests can be even stronger. Hence, vested interests can serve as a strong barrier to the solving of our social problems.

Legitimation

A third very powerful impediment to the solving of social problems is when people with the power and money have the ability to socially construct beliefs, laws, informal norms, and traditions in such a way as to make it appear that it is just and right for them to be and stay in power. In other words, if the people in power can convince the rest of us that the current situation is the way it should be, it is highly likely that the existing situation will continue and hence there will be no social change and no progress toward the solving of social problems.

These first three barriers of power, vested interests, and legitimation can serve to maintain the existing status quo with all of its attendant social problems for years—even hundreds of years. Although these three barriers are extremely powerful in maintaining the existing social conditions and therefore inhibit the solving of social problems, there are other barriers that we now discuss to let you be more fully aware of the challenges we face in solving our social problems.

Redefining the Social Problem

Another deterrent to solving our social problems has to do with how we define them. A group of people, such as a civil rights group or a group concerned about the environment, may define a social problem one way, whereas political authorities who deal with the social problem define it another way. Yet a third group, the people who carry out a social policy to solve a social problem (usually people in government bureaucracies), may define it a third way. Given these different definitions, these three groups can disagree on what social policy is best to solve a certain social problem.

Disagreeing With Social Policy

Even when various groups agree on the definition that something is a social problem, they can disagree on social policy. Different political philosophies, for example, will often result in different social policies. Let us use poverty as an example. Conservatives may suggest decreasing taxes on businesses so that these businesses will have more money to expand existing facilities, thereby creating more jobs. Liberals may suggest increasing social programs, such as job training and loans to get more education, thinking that if the poor get an opportunity to be trained or educated, they can fill new types of jobs that have higher incomes and chances for more upward mobility. Radicals, on the other hand, may assert that capitalism is the problem in that it does not provide enough jobs, does not provide enough jobs that pay above the poverty line, does not distribute goods and services somewhat equitably to all people, and hence does not address the needs of many people. Thus, radicals' solution is to change to a more socialistic system where everyone would be guaranteed a job at a decent income and all people's needs would be satisfied. As you can see, even though everyone might agree that there is a social problem of poverty, they may disagree on the social policies to use.

Why do people have different political views? There are a number of reasons. One reason relates back to what Durkheim (1933) suggested is the nature of modern society. That is, people in modern society have a greater variety of values and beliefs than did people in societies of the past. For example, a conservative may value individualism and freedom to make as much money as possible even though this may result in some people having millions of dollars while others are homeless and penniless. A liberal, on the other hand, may value having more equality in society where money and access to resources are distributed more evenly at the expense of losing some freedom to make as much money as is possible.[1] These different values and beliefs will cause people to have different social policies, making it difficult for everyone to settle on an agreed-on social policy that will solve a social problem.

Desiring a Minimum of Government

Some people, called *libertarians,* have a personal philosophy that there should be little or no government.[2] They believe that more government means less freedom for people. Even though they might agree that some social condition is a social problem, they do not want the government to help solve the problem because that would mean creating a larger government. They argue that other methods, such as more individual determination and responsibility, should be used to solve the social problem. Consequently, they will fight against nearly any expansion of government even though the use of government could be a major means to solve a number of social problems.

Although conservatives are not as extreme in their desire to limit the government as are libertarians, conservatives also believe that a minimum of government is best. When libertarians and conservatives combine to form a coalition, they can be a powerful force that restricts the use of the government as a means to solve social problems. As you go throughout your life, you will probably notice a trend in politics where libertarians and conservatives will work to limit or decrease the programs and services of the government, whereas liberals will work to retain or increase programs and services of the government. This trend is the result of two different philosophies; libertarians and conservatives will emphasize more freedom, potentially at the expense of having more inequality, whereas liberals will emphasize more equality, potentially at the expense of having less freedom.

As a side note, if you are taking a course in social problems or some course related to social problems and reading this book, you may want to think about your own political philosophy as it relates to what social policies you believe should be used to solve social problems. This is one of the things you may want to think about as you work your way through the course. If you find inconsistencies between your beliefs and the social policies you think will solve social problems, you may want to rethink your political philosophy. So, ask yourself the question, "Where do my social policies fit—as a libertarian, a conservative, a liberal, or a radical?"

Given that libertarians want to have a minimum of government and not use government and its resources to solve social problems, you might wonder how they would go about solving social problems of poverty, crime, homelessness, and so on. Their answer is that each person should be responsible for himself or herself, and therefore each person should rely on himself or herself to solve his or her own individual problems. They assert that if we all are responsible for ourselves, we will solve our social problems.

The libertarians have been criticized for this stance because a number of problems in our society are caused not by individual inadequacies but rather by the social structure. Some individuals are indeed able to solve the difficult situations they are in by themselves, but by and large, if some aspect of the social structure causes a social problem, most people experiencing the social problem will not be able to get out of their situations by merely taking more personal responsibility. For example, no matter how much people take more responsibility for themselves and their situations, this will not change the lack of enough jobs or the lack of enough decent-paying jobs in a country. In other words, to solve social problems, something more needs to be done than merely expecting individuals to take on more personal responsibility. In other words, various aspects of the existing social structure, such as more jobs and more decent-paying jobs, must change if we want to solve our social problems.

You may ask what I mean when I say the social structure causes a social problem. Maybe one way to show you this is to give the example of African Americans living in our country during the 1950s. There were laws and informal norms during the 1950s that did not allow African Americans to go to the same schools, eat in the same restaurants, stay in the same motels, swim in the same swimming pools, sit in the same movie theaters, live in the same neighborhoods, worship in the same churches, or join the same clubs as did whites. These laws and informal norms also did not allow African Americans to vote or hold various political offices. All of these beliefs, laws, informal norms, and customs that prohibited African Americans from doing these things combined to form what sociologists call a *social structure*, that is, social patterns that occur over and over and that, in this case, caused African Americans to have less opportunity, less upward mobility, and hence less income, wealth, power, and prestige.

To the degree that the social structure of a society limits the opportunities of groups of people and therefore limits their upward mobility (e.g., African Americans, Native Americans, Hispanic Americans, gays, women, the handicapped), we can say that the social problems that develop from such a lack of opportunity, upward mobility, and hence lack of money, power, and prestige are caused largely by the social structure rather than by the actions of individuals. In such a situation, to solve a social problem, it is the social structure, rather than individuals, that needs to change. For example, individual African Americans could be honest, hardworking, and responsible during the 1950s, but because they did not have the opportunities that whites had due to the existing social structure, their individual behavior had little or no effect on their life situations. The social structure, rather than their own individual efforts, largely determined what they could do.

Individual Change Versus Social Structure

An additional obstacle to the solving of social problems has to do with how much social policies emphasize changing individuals only. If the social policies applied are calling for individuals to change, the social problems will not be solved. Why? Because the real causes of the social problems are structural. One of the key things we need to do as we study social problems is to discover how the existing social structure causes social problems. If we are not aware of this connection, we are less likely to solve our social problems.

Values

The values that we espouse can be an impediment to solving our social problems. As hard as it may be to believe at first, our values might not always be consistent with other people's values, and this hinders the solving of certain social problems. Let us take the example of gender inequality. When asked, most people in our society would say that they believe in equal opportunity. Given the ideology taught to us and how we have been socialized, most Americans would say, "Yes, of course, I believe in equal opportunity."

Yet a number of people in our society take the Bible and what it says as the literal truth. Certain passages suggest that women are supposed to be subordinate to men. For example, in the Old Testament in the book of Leviticus (27:3–4), the Bible implies that women should be paid 60% of what men are paid. In the New Testament in the book of Ephesians (5:22–24), the Bible says, "Wives, submit yourselves unto your own husbands." A third passage is in I Peter (3:1 and 7), where wives are told to "be in subjection to your own husbands" and a wife is described as "the weaker vessel."

For a number of people in our society who take a literal interpretation of the Bible, the value of equal opportunity and the values of the Bible pose a dilemma such as the following:

> If I go by the society's value of equal opportunity, I will break biblical teaching by treating women equal to men. If I go by biblical teaching, I will break society's value of equal opportunity by treating women unequal to men. What should I do?

One of the consequences of these conflicts in values is that social problems are not solved. Instead of changing laws and informal norms and other aspects of the social structure quickly, these parts of the social structure are argued over extensively. As a result, social change comes about more slowly, with the result that social problems are solved more slowly.

This situation becomes especially difficult when the laws and informal norms in question are seen by some people as having been created by God. One can predict that whenever there is controversy over what people see as God's laws, there will be a long and emotional debate that will result in taking more time to create a social policy to solve the social problem, for example, in the areas of gender inequality, abortion, homosexuality, gay marriage, stem cell research, and teaching evolution.[3]

Existing Social Policies Versus New Social Policies

An additional obstruction is that there may be other policies that dealt with this problem previously that conflict with new policies. Peyrot (1984) noted, for example, that we have responded to drug abuse via both a criminal justice policy and a medical treatment policy. These two models do not mesh well to create a coordinated social policy toward drug abuse (p. 91). In other words, our discordant policies can work to slow down the process of solving a social problem.

Once a social policy is implemented, there is a time period when the new policy is implemented to see whether it works. In a number of situations, social policy did not work the way it was intended to work on paper. Unexpected things happened, there were not enough resources to solve the problem, and so on. When this happens, those who were skeptical of the policy now have new ammunition to fight it. It can be attacked, thereby decreasing enthusiasm for the policy and making more people question whether the policy will really solve the problem.

If the skeptics are powerful enough to make their doubts known to the policymakers, resources can be cut back or the social policy itself can be dismantled. A result of this process is that we do not solve the social problem. This does not mean that the policy was the wrong policy even though it may have had flaws. It does show the power of interest groups and how they can negate a social policy that could have solved, or at least ameliorated, a social problem.

Mass Media as a Business

The mass media can act as an obstruction to the solving of a social problem. For example, the mass media may be seen as seeking the truth. In many cases, this is true. However, we need to remember that the mass media, such as newspapers, radio stations, and television networks, are businesses that seek profit. Consequently, they may play up a social problem that will sell newspapers or increase listener or viewer ratings, and in so doing they may downplay another social problem. The result is that the less publicized social problem does not get the attention it needs for it to be addressed satisfactorily.

Lack of the Best Information

Even if social policymakers in state legislatures and Congress are genuine in their attempt to solve a social problem, they might not always have the best information on the topic with which to make the right decisions to solve the social problem. For example, policymakers might not read a sociological study that has important information on a social problem. Sociologists, like other academicians, report their research in professional journals that many times are read mainly by other academicians. This situation can create a closed loop of information. Academicians will read each other's published research and understand the implications of the research, but policymakers might not always be aware of such research. Hence, the research that could be crucial to the solving of a social problem goes unread by the policymakers. For example, in reference to a piece of sociological research, Wilson (1993) noted, "These important findings, buried in an academic journal, were apparently not discussed by the media and were probably ignored by policymakers" (p. 17). So long as the findings of sociological research and other social science research are not used in addressing our social problems, as a society we are more likely to create inappropriate social policy and are less likely to solve our social problems.

Recipients of Government Services

Another obstacle that works against the solving of a social problem is the fact that as recipients of government services receive these services that bring relief to their daily situations, they may define the social problem as being solved because they are getting help. The social problem, in reality, has not been solved; rather, it has only been ameliorated. The result may be to "cool out" and "cool down" the recipients and other people advocating for them (Piven & Cloward, 1971).

Nature of Congress

Finally, another obstruction to the solving of social problems has to do with the nature of Congress. Many times, Congress is more of a reactive institution than a proactive one. Typically, as a social problem occurs,

Congress reacts to address the problem rather than acting to prevent the potential social problem from materializing in the first place. As a result, members of Congress do not plan ahead and are not as visionary as they could be.[4] David S. Broder, a prominent newspaper editorialist, stated, "Our system of government is notoriously short-sighted; we do not act until a crisis is upon us" (Broder, 2002, p. A11).[5] Why is this the case? A key reason is that members of Congress know that they can be voted out of office rather quickly (every 2 years for a member of the House of Representatives) if they wander too far from their constituents' views. Hence, they end up being less bold, farsighted, and visionary and hence less likely to solve social problems. In other words, their primary concern is to hold onto power.[6] Schumpeter (1976) put it this way: "We must start from the competitive struggle for power and office and realize that the social function is fulfilled, as it were, incidentally—in the same sense as production is incidental to the making of profits" (p. 282).

As you can now see, there are many barriers to the solving of our social problems. You might be rather pessimistic at this point. However, there are also a number of possibilities that we need to consider to help us realize that not all is "doom and gloom." Let us look at these possibilities.

Possibilities for Solving Our Social Problems

Social Construction of Reality

We created this social world, and we can change it. Our ancestors created beliefs, values, customs, laws, norms, political and economic systems, and so on. Over hundreds or thousands of years, we, as humans, forget that we are the creators of these social phenomena. Once we realize that we created social phenomena, we are much more likely to realize that we can make various changes in our social phenomena. For example, we have gotten rid of slave systems and various kinds of prejudice and discrimination and have replaced these human creations with new human creations that provide more equal opportunity for people. So, the more we realize that we socially constructed this reality and that we can change it if we like, the more we are likely to take the next step of actually changing it. The more we realize this, the closer we will be to solving, or at least ameliorating, our social problems.

Addressing Social Problems

If you think about it, we have made progress with a number of our social problems. We have recognized different kinds of prejudice and discrimination (e.g., racial/ethnic, religious, gender, homosexual, handicapped, age) and have constructed new beliefs, laws, norms, and customs to decrease prejudice and discrimination. Our society is far from being perfect on this dimension, but compare our society today with that, for example, in the year 1950, when African Americans were not allowed to go to many schools, live in many neighborhoods, or have many kinds of jobs; homosexuals were not allowed to even express who they were and how they felt personally; women could not play sports and were not encouraged to go to college or to become doctors, lawyers, professors, ministers, or chief executive officers; and handicapped people could not make their way downtown, through a college campus, or to another floor in a building. All of these examples show that we have made changes, that more Americans are being taken into consideration, and that we are addressing our social problems.

Consciousness

More people in the United States and throughout the world are becoming conscious of what hurts them, what alienates them, and what holds them back. Computers, e-mail, the Internet, travel within a country and to other countries, study in other countries, the reading of books and newspapers, and the watching of television all have at least one thing in common, namely that they increase the consciousness of humans around the world, whether they intend to or not. News, knowledge, information, insight, new ways of thinking, and different ways of thinking are spread via computers, travel, the mass media, and so on. Whether we want consciousness about the world and the conditions of the world to spread or not is one thing. The fact is, regardless of whether we want such consciousness about world conditions to spread at what seems like a geometric rate, it is spreading quickly. Recall in our theory of conflict and social change and the accompanying causal model that a key factor in social change is the consciousness or awareness of existing social conditions and their inequalities. No doubt, as more people become conscious of these conditions, it is only a matter of time before they will form groups and organizations to address these concerns.

These first three possibilities—we socially constructed this social world, we can already see progress on a number of fronts, and we are increasingly conscious of the social conditions within which we live—are potentially powerful engines for social change and the eventual facing of our social problems at both the national and international levels. In addition to these three influential factors, there are other factors that work toward the solving or ameliorating of social problems.

Redefining

One way we can solve social problems is to redefine them as no longer being a problem. At first thought, this may sound ludicrous, but this was done when we repealed the prohibition amendment, thereby making alcohol legal to drink again. At one time, we defined drinking alcohol as a social problem and made it illegal. At another time, we ceased to define it as a social problem and made it legal.

This act of redefining has also gone in the other direction, that is, to label as a social problem a situation that, at one time, was not a social problem. Peyrot (1984), for example, noted how until 1875, "there were virtually no legal restrictions on the use of drugs such as opium, heroin, morphine, codeine, cocaine, and marijuana" (p. 86). He pointed out, however, that between 1875 and 1912, most cities and states began passing laws against these drugs out of "fear of minority-group drug users" (p. 86). Note that the act of redefining occurred over a period of time where much discussion and debate occurred, leading to the eventual social construction of new ideologies toward these drugs. As we might expect, with the changing of laws and ideologies came the relabeling of people who used and sold drugs from a good label to a bad one. As you can see, this redefining process is very political.

At first thought, we may think that redefining a social problem as no longer a social problem is the easiest way to solve a social problem. However, redefining a social problem is not that easy. A great deal of convincing needs to be done before some social problem gets redefined as no longer a social problem. For example, marijuana has been discussed and debated since the 1960s as to whether or not it should be legalized. Although various states have reduced their penalties on illegal marijuana use, they still have not come to the point of redefining it as legal. Any attempt to redefine a social problem will typically require the use of resources such as money, time, and people. So, redefining a social problem is not necessarily easy, quick, and/or inexpensive.

Concern by Powerful People

Another way a social problem can get attention and increase the possibility of getting solved is for it to become a great concern to powerful people whose vested interests are at stake. Once their interests are threatened, they not only will have the motivation to solve a social problem but also will have vast resources given their positions of power, access to influential networks, access to much money, and access to people working for them who can work full-time on the social problem. Thus, if powerful people get seriously involved in wanting to solve a social problem, the chance of solving the social problem becomes much greater.

Attention

Sometimes political authorities are in a social situation where they must pay attention to a social problem. This is shown most clearly by the news media and their influence. For example, the media can give a social problem much visibility and attention and not let the controversy of a social problem die before the reading and viewing public and political authorities. If the media champion a particular social problem and persist in covering it, that social problem is more likely to be addressed than are other social problems not discussed by the media. In a sense, if and when and for as long as the media take on a social problem, the powerful people of the society may have to face the social problem whether they want to or not.

A Matter of Degree

A fact we must accept in solving social problems is that many social problems will be solved by a matter of degree rather than in an all-or-nothing way. It would be nice if social problems got totally solved once they were addressed. That may happen at times, but the more realistic course of events is that because the solving of a social problem can conflict with the power, vested interests, and values of various groups of people, we should expect various compromises to occur. The result of these compromises is that many social problems will be ameliorated rather than solved completely.

Shared Values

Although conflicting values that people have can stop or slow down the solving of social problems, at times they can help to solve social problems if most of the members of society share the same values. There are some values in our society, such as equal opportunity, fairness, freedom, justice, and democracy, that are widely accepted among the American people.[7] These shared values can be a starting point in how we solve our social problems.

The importance of shared values in solving social problems can be shown in the civil rights movement of the 1950s and 1960s. Martin Luther King, Jr. applied the previously mentioned values to prompt whites and political leaders to reformulate laws. Before the 1950s, although African Americans believed in these values, they did not get to experience them in their daily lives. For example, African Americans could not attend the same public schools that whites attended, could not attend public universities such as the University of Alabama and the University of Mississippi, and could not join country clubs. In addition, they could not stay in the same motels or eat in the same restaurants as did whites, shop in certain stores, work in certain jobs, vote, or hold public office. The values of equal opportunity, fairness, freedom, justice, and democracy rang hollow for them. King and other civil rights leaders used this gap between our shared values and the existing laws to justify changing the laws in the direction of these values.

By King's pointing out the discrepancy between our shared values and our laws, our society began to make changes to solve the social problem of racial prejudice and discrimination. Laws were enacted so that African Americans could attend public schools and universities, play sports at southern universities (Eitzen & Sage, 1997, p. 263),[8] live in dorms on campus rather than live off campus in private homes and need to commute to the university,[9] vote in elections, run for and hold public office, live in any neighborhood they could afford, and work at any job they were capable of doing. The changes in these laws allowed African Americans not only to believe in these values but also to experience these values.

Is everything perfect, and is there no more racial prejudice or discrimination? No. We still have work to do. However, sharing values in common and working to see that our laws are consistent with our values is a step in the direction of solving the social problem of racial prejudice and discrimination. As you can see, values, especially values that are widely shared, have the potential to play a key role in solving our social problems.[10]

Coordination of Social Policies

Another key area in helping us to solve our social problems is how social policy is carried out. If a new social policy is coordinated with other existing social policies, the social problem has a better chance of being solved. Schoenfeld, Meier, and Griffin (1979) and Peyrot (1984) discussed the situation of new social policies coming into conflict with existing social policies, with the result that the social problem is less likely to be solved or more likely to be solved at a slower rate.

Consequently, one crucial aspect in creating new social policy is to study how new and existing social policies can relate to and complement each other. If there are potential points of conflict between new and existing policies, how can we smooth over these rough spots before we carry out the new policy? If we do not answer this question first, we build into the new social policy a predisposition for failure or at least for a slower pace at solving the problem. So, to solve or at least ameliorate social problems, we need to think ahead and coordinate new and existing social policies.

Valid Data

We are more likely to solve our social problems when political authorities draw on valid data as they consider what social policies to create. This kind of data is collected by sociologists and other social scientists. For example, in sociology we create research designs to arrive at the best way to collect data on a social topic. We then use various methods of gathering data such as the survey, the interview, and/or participant observation. Once the data are collected, we analyze them to see what conclusions we can make. Once we know what the data are saying, we can inform political officials of our findings. The political officials in turn can use this information to make more informed social policy. In short, the use of the best data available will give policymakers a better idea of how to solve our social problems.

Incentives Versus Disincentives

Another key factor in solving social problems is for there to be more incentives to solve the problem than incentives to do little or nothing about the problem. As many groups begin to see incentives to solve a certain social problem, they are more likely to join in the support of creating a certain social policy. As these incentives are made known such as through the mass media, there is an increasing probability that something will be done about the social problem.

Related to the idea of incentives and pointing them out to people as a means of solving a social problem is how these two actions relate to exchange theory (see Chapter 1).[11] As we discussed previously, social problems are not as likely to be solved whenever people believe they are expected to give up resources and yet receive nothing in return. However, if people are shown how they can benefit from giving up some of their resources to solve a social problem, they will be more willing to part with some of their resources in the form of taxes. For example, if they believe that paying more taxes will decrease crime, and that this will make the streets safer for them and their families, they may be more willing to pay more taxes. In other words, part of the process of solving social problems will be to help people realize "what's in it for me."

Perceived Urgency

A social problem is more likely to be solved when political authorities and the general public begin to see the urgency in addressing the problem, that is, when there is a feeling that something must be done and must be done *now*. Such a state of urgency can act as a stimulus to address a social problem. We observed this phenomenon after the tragedy of September 11, 2001, and how our government took a number of steps to beef up security at our airports.

Many People Affected

In some social problems, relatively few people are negatively affected. In other social problems, many people are negatively affected and many others feel the negative effects indirectly. When many people are affected either directly or indirectly, we are more likely to address a social problem. Hence, the more people are affected by a problem, the more likely society will address that problem.

As you can see, there are a number of situations that can lead to conditions conducive to solving our social problems. So, even though there are a number of barriers to solving our social problems, there are also a number of other factors that push us in the direction of solving our social problems.

How Can Sociology Help to Solve Social Problems?

There are a number of ways in which we, as sociologists, can contribute to solving our social problems. Let us discuss these various ways in turn.

Consciousness-Raising Role

One way sociologists can help to solve social problems is to analyze social phenomena to arrive at what we think may be new social problems arising in the society and to inform the media and political authorities of our findings. In a sense, we can serve a consciousness-raising role by pointing out what we see as the social problems of our times (Wilson, 1993). We can be guided by the criteria suggested in Chapter 1, that is, the number of people being affected and whether or not the social conditions endanger lives.

We can also carry out this consciousness-raising role when we teach and publish books and articles about social problems. Our teaching and publishing activities can serve to legitimize certain social conditions as social problems in the eyes of the public. We can also use the theory of conflict and social change and causal model discussed in Chapter 1 to help people become more conscious of the key variables that are a part of the process of a social problem.

Research

Sociologists can help to solve social problems through the research we do. For example, we can pinpoint causes, locate what stage the social problem is in, uncover the values and vested interests that are in conflict, and analyze to what degree current social policy has an effect on the solving of the social problem. Kingdon (1993) noted, "Social scientists can be very good at documenting the existence, frequency, incidence, and intensity of a condition" (p. 48). As a result, our analysis will provide a clearer understanding of the social problem for the general public and for political officials. For example, former Senator Bill Bradley of New Jersey told William Julius Wilson, a sociology professor, that Wilson's (1993) book, *The Truly Disadvantaged*, "illuminated their [the senators'] understanding of the problems of ghetto poverty, raised their consciousness, and increased their awareness of the need for effective public policy to address these problems" (p. 9).

"If, Then" Statements

Sociologists can also help to solve social problems by making what Berger and Kellner (1981) called "if, then" statements (p. 76). Berger and Kellner asked the following: *If* certain social conditions exist, *then* what can be done? For example, *if* in capitalism there is unemployment and there are jobs that pay below the poverty line, *then* what can we do to solve poverty? In other words, sociologists can specify the given social conditions and proceed to create the various choices of social action that can be taken within those conditions. By creating "if, then" statements for social problems, sociologists can get a clearer idea of the possible social policies that can be attempted and the possible consequences for each social policy. Such knowledge will prove to be invaluable for those who create social policy.

Analyzing Social Policy

Another way sociologists can contribute to the solving of social problems is to study the various social policies currently addressing a particular social problem, see how they are related, and suggest how they can be further interrelated to do a better job of solving the social problem. Given that different social policies are created at different points in time under different political administrations with different political philosophies, such an array of social policies addressing a specific social problem can consist of duplication in some areas, big holes in other areas, and working at cross-purposes in still other areas. When they know these things about the current social policies, sociologists can analyze these policies and suggest possible ways they could be more fully integrated.

Sociologists can also study the potential impact of a new social policy on a social problem. We can see whether the policy is doing what it is supposed to be doing. We can assess how successful the new policy is at solving or decreasing the problem and what the unintended consequences of this new policy are. With such data in hand, political authorities will be better able to decide how well a new social policy is doing and make adjustments to the social policy if necessary.

Addressing Which Problem First

It appears that some social problems are the key to solving, or at least lessening the severity of, other social problems. For example, it seems that solving poverty will help to diminish other social problems such as crime, spouse and child abuse, divorce, drug abuse, and poor health. That is, poverty is a key independent variable that contributes to these other social problems. Although we do not assert that the solution to poverty is the panacea for all of these other social problems, we do predict a decrease in the rates of these other social problems.

Because poverty exacerbates other social problems, it follows that one of society's primary concerns could be the focus on poverty. One of the key factors that could solve the problem of poverty is to offer a sufficient number of decent-paying jobs, that is, jobs that pay above the poverty line. Consequently, social policies that increase the number of decent-paying jobs would be at the top of the list of which policies political officials might want to address first.

Hence, it follows that it would be very beneficial to study how social problems interrelate. Such analysis will provide us with a clearer idea of which social policy will give us "the biggest bang for the buck," that is, how it could address two or more social problems. Armed with such knowledge, we can then choose certain social policies that will use our limited resources most effectively. Sociologists can therefore contribute to this effort by analyzing the various interrelationships among social problems.

Studying the Social Policies of Other Countries

Sociologists can also study the social policies of other countries to learn how these policies might address our social problems. For example, we can find out what specific policy was attempted, what resources were used, what the unintended consequences of the social policy were, and how the other country is similar to or different from our country; for example, we would need to take into consideration the population of the other country and the degree of its people's heterogeneity such as race, ethnicity, and religion. For example, let us say that we wanted to study the country of Sweden and how it has addressed the social problem of poverty. We would need to keep in mind that Sweden is a much smaller country with a more homogeneous population compared with the United States. With these differences in mind, we could assess Swedish social policy to see how suitable it would be for our country. In other words, sociologists could collect comparative information on the social policies of various countries as a means of having a larger pool of information to draw from in deciding the kind of social policy we might try in the United States (Eitzen & Leedham, 1998, 2001).

The Victims of Social Problems

Sociologists can also collect data on the victims of a social problem. For example, we can find out the answers to the following questions. What are the victims experiencing day to day? How are they coping? What is happening to them that they believe the rest of society does not realize? What do they think needs to be done to solve this social problem? These are some of the questions that sociologists can answer by doing research on victims. What we find out from this research can guide political authorities in the development of social policy to address these social problems.

A Sociological Perspective

Sociologists can help others to develop a sociological perspective with which to view social problems and their solutions. Policymakers are already using sociological theory, hypotheses, concepts, and research findings to analyze social problems. Weiss (1993) asserted,

> Evidence from social science research can reduce disagreements over matters of fact (e.g., whether fewer pregnant women are receiving prenatal care, whether vocational education improves employability and job performance). In doing

so, it helps to raise the level of debate, freeing policy actors to talk about matters of value—which is their proper province. Analogously, the concepts and theories of sociology make a difference. They are helping to make public decision makers more sophisticated about social structure and group processes (less content with individual-level explanations for social phenomena), and they are gradually infusing political thinking with more complex and subtle notions of conflict, social disorganization, community norms, social movements, and other sociological constructs. (p. 29)

Weiss (1993) urged that sociologists undertake two actions as a means of solving social problems: (a) make a sustained effort to reach policymakers in terms of disseminating sociological findings and conversing with policymakers and (b) share the sociological perspective with policymakers so that they can gain a much greater "understanding of the forces and currents that shape events" in our society (p. 37).

No Crystal Ball

As you can see, sociologists can contribute in many ways to the solving of our social problems. Yet we do not have a crystal ball. That is, we cannot say for certain that if a certain social policy is attempted, it will solve the social problem. There are three reasons for this. First, social life is complex in that there are a number of independent variables that influence every social problem. Second, we do not always know precisely to what degree each independent variable influences a social problem. Third, we do not always know all of the independent variables causing a social problem.

On the other hand, sociologists know many of the independent variables that influence social problems. Much research has been done to give us good ideas about these variables and their influences. We have already collected immense amounts of data and discovered many interrelationships. We are continually learning more and more about social phenomena in terms of how certain independent variables influence social phenomena. Consequently, we have a clearer and more comprehensive picture of what is occurring in social life than ever before in human history.

Because sociologists have gathered considerable data on social problems, because they have discovered many social patterns within social problems, and because they have isolated many variables that cause social problems, it is evident that sociology not only can but also must play an important role in solving our social problems. Beginning with the next chapter, I share with you how sociologists and the sociological perspective can be used to solve our social problems.

Questions for Discussion

1. What might be some things we could do to overcome the problem of conflicting vested interests in a social problem?

2. How could Congress become a more visionary body and plan ahead instead of reacting to what goes wrong?

3. When should we consider a social problem to be solved?

4. There are different political orientations that want different social policies to solve a social problem. What are some steps we can take to resolve these differences?

5. How can we be more confident that the social policies we use will work?

6. What other values besides equal opportunity, fairness, freedom, justice, and democracy do you think we share in common in the United States that could serve as a starting point to solve social problems? In what way?

7. Which political orientation do you think is the best for solving our social problems? Why?

8. Which political orientation do you think will be used the most during the next 10 years to address our social problems? What is your reasoning?

9. Should sociologists take a particular stance on a certain social policy?

10. Given their knowledge about social problems, should more sociologists run for political office? Why or why not?

3

How Can We Solve the Problem of Increasing Inequality?

One of the biggest social problems that we are having in the United States, as well as around the world, is the problem of growing inequality. In this chapter, I define inequality, point out statistics on it and how it is growing, note some of the key causes and consequences of it, and suggest what we can do about it. Do we want to continue to increase inequality? Do we want to maintain the current amount of inequality? Do we want to decrease inequality in the United States and in the world? As you read this chapter, keep these three questions in mind and see what you believe we should do.

Definition and Statistics

Inequality in a society occurs when people have differing amounts of money (yearly income and/or wealth), power, and/or prestige.[1] I need to define some of these terms so that we all will be thinking the same things as the terms are used throughout this book. The kind of money that we call *income* is money that we get from the job we have. *Wealth,* the other kind of money, represents what we own. For example, if you own a home and stock, you have some wealth. Some people have a lot of wealth because they own a number of homes and have millions of dollars worth of stock, whereas other people do not own a home and have no stock. As you know, the variation in income and wealth is huge. I define the term *power* by using the definition of the great German sociologist of the early 20th century, Max Weber: "We understand by 'power' the chance of a man or a number of men to realize their own will in a social action even against the resistance of others who are participating in the action" (Weber, 1968, p. 926). Simply put, power is the ability to make people do things even against their will. If someone is holding a gun on you, he or she can take your money. If someone has the legal ability to make people do things against their will, I define this as *authority*. For example, a police officer has the authority to give you or me a speeding ticket even though we do not want such a ticket. By *prestige,* which is sometimes referred to as *status,* I again use Weber's (1968) definition: "We wish to designate as status situation every typical component of the life of men that is determined by a specific, positive or negative, social estimation of honor" (p. 932). In other words, when we give people high honor or high respect, we give them prestige. For example, we tend to give higher prestige to medical doctors and U.S. Supreme Court justices and to give lower prestige to garbage collectors and custodians.

In sociology, we call each of these a *dimension of inequality* and gather data on how much money (yearly income and/or wealth that people own), power, and prestige people have. For example, with regard to money inequality, some people in our country make incomes at the minimum wage level and earn only $5.15 per hour or $206 per week and $10,712 per year. Other people, such as teachers, carpenters, and plumbers, make yearly incomes in the range of $30,000 to $60,000. Medical doctors, dentists, lawyers, and some businesspersons tend to make from $200,000 up into the millions of dollars per year. Some people in our country make billions of dollars per year.

Not only is there substantial inequality in our country, but statistics suggest that it is growing. Let us look at money inequality—both income and wealth—and see how it has been on the rise. With regard to wealth inequality (again, with wealth being what people own in the form of homes, stocks, bonds, land, and buildings), 1% of the richest Americans owned more of the country's wealth (38.5%) in 1995 than in any year since 1922 when data were first gathered (Kerbo, 2003, p. 35). If we look at stock ownership, we find that 1% of Americans owned 46.7% of all stock in the United States in 1989 but owned 51.4% of the stock by 1995 (p. 34). Moreover, 10% of Americans owned 83.8% of the stock in 1989 but owned 88.4% of the stock by 1995 (p. 34). So, these statistics suggest that there has been growing wealth inequality in our country.

As for income inequality, the richest 20% of people made 43.3% of all the income in our country in 1970. By 1999, they made 49.4% of all the income (Eitzen & Zinn, 2003, p. 36). At the same time, the poorest 20% of people earned 4.1% of all the income in 1970 but earned only 3.6% by 1999 (p. 36). In a study of rising income from 1972 to 2000, while the income of Americans in the 90th percentile of income rose 34%, the income of Americans in the 99.99th percentile of income rose 497%—nearly 15 times faster (Krugman, 2006b, p. A7). Economist Paul Krugman reported that data for 2004 show that "a small fraction of the population got much, much richer" (Krugman, 2006a, p. A9). The recent Bush tax cuts mean that middle-income people will receive a 2.3% increase in their incomes after taxes, whereas rich people (those earning more than $1 million per year) will receive a 7.3% increase in their incomes after taxes, thereby creating more income inequality. So, these statistics suggest that there is also grow-ing income inequality in our country.

As money inequality grows, this usually means that those with more money will have more power because they can use their money in many ways to get what they want and carry out their own vested interests, some-times at the expense of other social classes (e.g., run for political office, give large contributions to people who believe as they do, threaten to give or not give to charities and various organizations unless these charities and orga-nizations conform to their beliefs). Consequently, those who have more money and power will have disproportionate influences in a society. They will decide whether factories in communities move or do not move, and they will have great influence in deciding the economies of communities (e.g., unemployment rate, poverty rate, homelessness rate, crime rate, rate of spending for public schools). They will have more influence in proposing bills in a state legislature or in Congress and will have more money and influential networks to lobby for the bills they want passed for their vested interests and possibly against the vested interests of other social classes. So, having more money (income and wealth) not only means that they have a higher standard of living for their families such as bigger houses, newer cars, and more expensive vacations, but it also means that they can have more political influence.

Causes of Growing Inequality

Most people in our country earn income from their jobs, and that is their only way of gaining money. Some people, on the other hand, not only make money from their jobs but also make money from the wealth they own. What seems to be one cause of growing inequality in our country is that people with higher incomes tend to get bigger raises than do people with lower incomes and that people who own stock, buildings, and land can get money from these investments (e.g., dividends, rent), whereas most other people in our country gain income solely from their jobs. Moreover, millions of people who gain income solely from their jobs have minimum wage jobs and do not receive raises each year even though inflation is going up each year, eating away what little money they do receive (the minimum wage is substantially below the poverty line for a family of four). The reason for this situation is that Congress decides when to increase the minimum wage. As a result of all these factors, there is growing inequality between people who have high incomes and much wealth and people who have moderate to low incomes and no wealth.

As people with high incomes and wealth accumulate more money each year compared with people who have low to moderate incomes and no wealth, those with higher incomes and wealth are also able to increase their power.[2] For example, when someone has a lot of money, he or she can give more contributions to elected officials in the hope of receiving favorable legislation in return. People with more money can afford to run for political office and spend more money than other candidates to win the office. People with higher incomes and wealth will also have jobs that allow them to interact with other people who have high incomes and wealth and also much power.[3] For example, a person who is a president or vice president of one large corporation, say one of the 100 largest corporations in our country (e.g., Wal-Mart, Exxon, General Electric, IBM, AT&T), many times not only will know people in the top positions in the other large corporations but also will serve on the boards of these corporations.[4] Assuming that 35 people are on a corporate board, there are 35×100 of the largest corporations or 3,500 people who control 75% of all the industrial assets in the United States (Kerbo, 2003, p. 189). Moreover, because a number of these 3,500 people are on one, two, or three other boards, there could be as few as 2,000 people who are the "movers and shakers" of much of our economy. In other words, as these largest 100 corporations become ever larger, they will also wield more power in our society as well as in other countries where they have corporate investments and therefore vested interests.

When these powerful corporate leaders speak, the rest of us listen and are influenced by what they decide to do. For example, if the board members of one of these corporations decide to cut thousands of jobs and close a number of plants, not only will workers in those plants lose their jobs but also the communities in which those plants are located will be hurt. Stores in those communities may close or suffer great financial losses; schools might not have enough money for teacher salaries, up-to-date equipment, and/or new buildings; and city governments may lose tax revenue that would have gone to pay the salaries of the police and fire department members and workers who maintain the streets and sewage and water systems.[5] So, a decision made by board members at corporate headquarters in a distant city can devastate workers and their communities (Mills, 1959b).

As you know, in the United States, we have a capitalistic economy. This means that some people are able to own factories and businesses where they can make a lot of money compared with people who are employees. It is in the nature of capitalism that employees will get wages (paid by the hour) or salaries (paid weekly, biweekly, or monthly), whereas owners get profits.

Also, the ideology in a capitalistic culture typically asserts that everyone is responsible for himself or herself and that there is much emphasis on the individual, with the individual getting ahead and surviving in a "dog-eat-dog" world. This ideology tends to justify or legitimize the idea of some people making more money than others and therefore justifies inequality and increasing amounts of it. This ideology is socialized into many people in our country from the time they are born, so that it seems like human nature to them to be individualistic and look out for themselves. As a result, the ideology we are socialized to believe in promotes the acceptance and legitimacy of a lot of inequality.

Another important cause of growing inequality in a capitalistic society is the belief that the influence of government should be kept to a minimum. If the government is kept to a minimum (e.g., providing only for the national defense), there does not need to be as much tax that people need to pay. With less tax to pay, wealthy people can keep more of their wealth, thereby increasing inequality. Also, with less tax revenue going to the government, the government has no choice but to provide fewer services, such as health care (see Chapter 10) and subsidized child care (see Chapter 4), thereby increasing inequality. So, the ideology of limited government tends to translate into more inequality in our country; the wealthy keep more of their money, whereas people with low to moderate incomes receive fewer services such as food stamps, medical care, and student loans.

Another cause of increasing inequality is the degree to which wealthier people can influence the government to decrease the progressive income tax. A progressive income tax is where people with higher incomes pay a higher proportion of their incomes in taxes than do people with lower incomes. If the tax system becomes less progressive, wealthier people pay less in tax, allowing them to keep more of their money and thereby increasing inequality. Because wealthier people typically have more money and connections with which to influence public officials than do people who are not wealthy, wealthier people can have more influence with respect to how people are taxed. They do not have total influence, but they will have more influence than the average citizen in our country.[6] Hence, it is in the vested interests of wealthier people to have less government that provides fewer services so that fewer taxes will be needed, thereby allowing for and justifying a lower progressive income tax. The bottom line is that with a lower progressive income tax, wealthier people keep more of their money, and this causes more inequality.

Other kinds of taxes can also increase inequality in a society. For example, because there is less tax on people who inherit wealth, the people who inherit wealth will be able to keep more of it, resulting in greater inequality. President George W. Bush has sought to abolish the inheritance tax, and doing so would work to the vested interests of wealthier people and result in increasing inequality.

The same idea can be applied to the sales tax. A sales tax is a regressive tax, meaning that lower income people pay a higher percentage of their incomes in taxes than do higher income people. For example, a higher sales tax on products that everyone needs (e.g., soap, shampoo, shaving cream, cars) will mean that people with low to moderate incomes will pay a larger percentage of their incomes in sales tax for these products than will people with higher incomes. Hence, as the sales tax goes up, inequality will also go up.

Another cause of growing inequality is the degree to which people of low to moderate incomes working for industries and businesses cannot organize into unions to protect their financial interests. The less these people are able to join unions to seek higher pay, the more likely it is that those at the top of these organizations will be able to keep a larger proportion of the profits for themselves. For example, German workers are the highest paid among the seven most highly industrialized nations, whereas American workers are the second to lowest paid (Kerbo, 2003, p. 28). A key factor in the difference of pay is the strength of unions in each country. Labor unions are strong in Germany, whereas they are relatively weak in the United States. In fact, in Germany, it is legally mandated that employees make up one half of the board of directors in a company (p. 510). This one difference alone gives German workers much more power compared with American workers.

A problem for workers in the United States that has hurt the amount of unionization and power of unions is the process of corporations moving their plants outside of the United States (known as *deindustrialization*) to other countries that allow for lower labor costs in the form of lower wages, no retirement benefits, and no health care benefits. Paying lower wages, providing no retirement benefits, and giving no health benefits together save a great deal of money for these corporations. They are able to make more profits, give higher salaries to their top officers, and give larger dividends to their shareholders. At the same time, the American workers who work at these plants can lose their jobs, health coverage, and retirement benefits, thereby resulting in more inequality.

For much of our country's history, prejudice and discrimination of different kinds, such as racial/ethnic, gender, sexual orientation, religious, age, and handicapped, have hurt people. They were hurt because when they were discriminated against, they were not given the same opportunities as were others. As a consequence, they typically needed to settle for lower income jobs with less power and prestige. This has been a major cause of inequality throughout the history of our country.

We have been reducing these kinds of prejudice and discrimination. So, over time, this cause of growing inequality should recede as we, as a country and as individuals, work to eliminate the remaining elements of prejudice and discrimination in our country and in ourselves. The current situation is not perfect, but we are headed in the direction of declining prejudice and discrimination. As this process continues, we should find that this factor will be one cause of growing inequality that will become less and less influential.

Consequences of Growing Inequality

One consequence of growing inequality is that people in the higher social classes will be more socially and physically distant from people in the lower social classes. That is, those in the higher social classes will be more likely to live in different neighborhoods, go to different public schools or attend private schools, attend different places of worship, and so on. With rising inequality, the higher social classes, by being more socially and geographically distant, will probably be less understanding of the members of the lower social classes. This situation increases the possibility that certain ideologies, such as "the reason why the poor are poor is because they are lazy," will continue and will be used to legitimize existing inequality.

Another consequence of growing inequality is that there will be a growing number of opportunities for the higher social classes compared with those for the lower social classes. For example, for the higher social classes, there will be more chances for travel throughout the world, for more education and higher quality education, for more consumer goods, for more and better health care, and for better retirement lifestyles. Unless some outside source such as the government intervenes by providing lower classes with a number of opportunities that the society, in its normal functioning within a capitalistic system, does not provide, the gap will continue to widen between the higher and lower social classes.

This process of a widening number of opportunities can lead to what sociologists call *feelings of relative deprivation,* where people in the lower classes will compare their situations with those of the people in the higher classes and feel deprived as well as resentful. This situation is especially likely to occur if the society socializes people that there should be equal opportunity in life but in reality there is not (refer to our theory of conflict and social change causal model in Chapter 1). If people feel relatively deprived and resentful, they may begin to question the legitimacy of the existing social conditions.[7] This can lead to riots and various outbursts of frustration from the lower classes, and these in turn can lead to less stability in society.

Options We Have With Inequality

In a society, we can increase inequality, keep the existing inequality, or decrease inequality. Let us discuss each of these and consider its implications. As we discuss these three options, think about what you believe we should do in society.

Increase Inequality

The first option is to increase our inequality even more. If we wish to increase inequality, we can do this by taxing the poor, working, and middle classes more and taxing the rich less. For example, we can increase the federal income tax on the poor, working, and middle classes and decrease the federal income tax on the rich. We can also increase the sales tax given that this will hurt people in the poor, working, and middle classes more than it will rich people because everyone needs to buy similar amounts of certain products such as soap, toothpaste, toothbrushes, toilet paper, shaving cream, razors, and shampoo. Poor people pay the same prices for these products and pay the same sales tax as do rich people. Consequently, when we raise the sales tax in a state or city, poor people will be poorer relative to rich people, thereby creating greater inequality.

Another way we can increase inequality in our country is to abolish the inheritance tax given that this will allow rich people to keep more of their inheritances, thereby creating more inequality. Currently, when people inherit the wealth from their parents when their last parent dies, the first $1 million is not taxed. However, any wealth above this amount is taxed at a graduated rate from 18% to 49%. Because people in the poor, working, and middle classes do not inherit very much money (if any), most Americans do not pay any inheritance taxes. Only people who receive very large inheritances will pay taxes on the amount above $1 million (personal communication from C. Campbell, tax accountant, March 3, 2004). If this tax were abolished, the people receiving these inheritances would keep more of their wealth, thereby creating more inequality in our country.

Another way to increase inequality is to decrease or abolish taxes on dividends from stocks and to decrease or abolish the capital gains tax on stocks, thereby allowing those who own stock and are making money from stock to keep more of their money. Given that 10% of Americans owned 88.4% of all the stock in the country in 1995, people in this group will be able to keep more of their wealth, with the result that inequality will increase in our country.

In addition to changing taxes that create more inequality, we can create more inequality by decreasing or abolishing social services that intend to help the poor, working, and middle classes survive or live better lifestyles. For example, if state legislators, governors, members of Congress, and the president decrease social services such as social security, decrease child care subsidies for mothers who are getting off welfare to work, decrease college loans and grants, decrease money for public schools, decrease Section 8 subsidized housing, decrease money for Head Start and/or Upward Bound programs, decrease money for unemployment compensation, decrease money for health care, and so on, the poor, working, and middle classes will be poorer, with the result that inequality will increase in our society.

So, increasing taxes on the poor, working, and middle classes, decreasing taxes on the rich, and decreasing social services are three ways we can increase inequality in our society. Many people would find these actions to be extremely distasteful given that the poor and working classes will have a harder time getting by each day.

When I have given an anonymous survey in my social problems class and asked my students whether they think we should increase inequality, keep it the way it is now, or decrease inequality, no one has ever voted to increase inequality. This does not mean that there are not Americans who do not want more inequality, but it does suggest that when students discuss the consequences of increasing inequality, some students opt for keeping it the same while most students vote to decrease it.[8]

Maintain the Current Amount of Inequality

The second option is to maintain the current inequality in our country with a certain combination of taxes and social services. Some of my students have voted for this option. Probably a number of people in our country, without any discussion over this issue, would vote for maintaining the current amount of inequality. However, given what I have observed in my social problems classes, if Americans discuss this issue and realize the negative consequences of rising inequality or maintaining the current amount of inequality in our country, I predict that the majority of Americans would, like the majority of my students, vote to decrease the amount of inequality in our country.

Decrease Inequality

The third option is to decrease inequality in our society. If we, as a society, choose to decrease inequality, we can do a number of things. We can decrease various kinds of taxes on the poor, working, and middle classes and, at the same time, increase taxes on rich people. For example, we can make the federal income tax more progressive, where the poor, working, and middle classes pay a lower percentage of tax and the rich pay a higher percentage of tax. Besides taxing income, we could tax the wealth of the rich, and this would also decrease inequality.

Another kind of tax we could change is the social security tax. Right now, only the first $94,200 that Americans earn each year is taxed at 6.2%. People no longer pay social security taxes on income they make above $94,200. So, a poor person making a minimum wage of $5.15 per hour, or $10,712 per year, will pay 6.2% in social security tax. At the same time, someone who earns $200,000 per year will pay only 2.9% of his or her income in social security tax (6.2% × $94,200 = $5,840.40 / $200,000 = 2.9%). A rich businessperson, pro athlete, rock star, or movie star making $10 million per year will also pay $5,840.40 divided by $10,000,000 = .0006% of his or her income in social security tax; in other words, a lot less than 1% of rich people's incomes will go to social security tax, whereas poor people will pay 6.2% of their incomes in social security tax. It is hard to believe that rich people pay less than 1% of their incomes in social security tax while people in the poor, working, and middle classes pay 6.2% of their incomes in social security tax, but that is the way it is. Hence, if we want to decrease inequality, we can have rich people pay more in social security tax by not having a limit on how much social security tax they pay while having people in the poor, working, and middle classes pay a lower rate. By the way, having the rich pay more in social security tax would also help to provide enough social security income for our elderly in the future, thereby going a long way toward solving our social security problem. Consequently, by increasing the social security tax on the rich, we could work to solve two problems: (a) decreasing our inequality and (b) making the social security system more solvent for our future elderly.

We can decrease the sales tax, making it not so hard on the poor to buy everyday products to survive day-to-day, and at the same time depend more on progressive taxes of various kinds such as federal and state income taxes and federal social security taxes. That way, more of our taxes would be structured so that the ability to pay taxes is tied to one's income and wealth—the more income and wealth, the more taxes people pay; the less income and wealth, the less taxes people pay.

Another way we can decrease inequality is to increase social services such as the following: increase food stamps for poor people; raise the minimum wage; increase social security for people in the poor, working, and middle classes; increase unemployment compensation; create more college grants and loans for people in the poor, working, and middle classes; create more child care subsidies for lower income single parents who are working at (near) minimum wage jobs so that they can work and survive at these kinds of jobs; create more housing subsidies for poor and lower income families; expand Head Start and Upward Bound programs for poorer and lower income people; and increase funding for public schools located in poor and lower income neighborhoods so that children from these neighborhoods get the same quality public education as do children in middle-class and upper middle-class neighborhoods.

In other words, taxing the poor less, taxing the rich more, and providing more social services for the poor, working, and middle social classes are three ways to decrease the inequality we have in our country. Diagrammatically, here are the three ways we could decrease inequality:

1. Increase taxes on the rich:

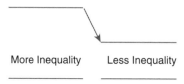

2. Decrease taxes on the poor, working, and middle classes and increase social services to them:

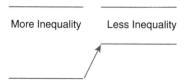

3. Combine Points 1 and 2—Increase taxes on the rich; decrease taxes on the poor, working, and middle classes; and increase social services for the poor, working, and middle classes:

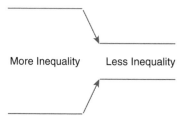

What Should We Do?

What should we do? This is a question that neither sociology nor any other social science (e.g., economics, political science, history, social psychology, anthropology, communications) can answer because science cannot tell us what we should do. Science, whether natural or social, can tell us many things. For example, it can give us the statistics on the unemployment rate, poverty rate, and homelessness rate. It can tell us what causes something to happen such as what causes unemployment, poverty, and homelessness. It can tell us consequences such as that more unemployment can lead to more poverty, homelessness, family stress, and crime and less tax revenues to pay for more police, courts, and prisons. It can also predict what may happen in the future. For example, given the racial prejudice and discrimination of the past 300 years in the United States, we can predict that African Americans will earn less income and have less wealth than will white Americans.

What science cannot do is tell us what we should do. The closest that science can come to answering a "should" question is to make "if, then" statements (Berger & Kellner, 1981). That is, if the society wants something to happen, such as a decrease in inequality, then sociology, along with other social sciences, can say what things can be done to achieve that goal.

In other words, sociology cannot tell us what we should do with regard to rising inequality. However, if the society wants to move in a certain direction, such as to decrease inequality, then sociology can help us to understand how we can do this and what the consequences might be.

What do you think? Should we increase inequality? Should we keep it where it is now? Should we decrease inequality? We have created the inequality that we currently have. Because we, as humans, socially construct the kind and amount of inequality that we have, we can change both the kind and amount of inequality that we have if we want to do so. We do not need to accept the inequality we currently have. It is up to us to decide how much inequality we want.

Questions for Discussion

1. Should we try to decrease inequality, or should we let it grow as it is currently doing?

2. Should we increase taxes on the wealthy as a way to decrease inequality?

3. Should we decrease taxes on the poor, working, and middle classes as a way to decrease inequality?

4. Should we provide more social services for the poor, working, and middle classes as a way to decrease inequality?

5. Where do you think inequality will go during the next 10 years—higher, lower, or stay where it is now? Why?

6. What are other ways we could decrease inequality?

7. How does having a capitalistic economy affect inequality?

8. Should the government be more involved in decreasing inequality?

9. How does the amount of inequality we have affect the social problems we have?

10. What do most Americans think about inequality? Why?

4

How Can We Solve
the Problem of Poverty?

In this chapter, our focus is on how we can get rid of poverty in the United States. By getting rid of poverty, we mean getting all poor Americans above what the federal government says is the poverty line (U.S. Department of Health and Human Services, n.d.). For example, the 2006 poverty line for a family of three (e.g., a mother and two children) was $15,927, and the 2006 poverty line for a family of four was $20,066 (Tax Policy Center, n.d.). So, the key question of this chapter is as follows: What can we do in our country to get all poor Americans out of their poverty or, more specifically, above the poverty line? Let us first make some assumptions and then discuss what we can do.

Assumptions We Can Make About Solving Poverty

Before we begin to discuss what we can do to solve the problem of poverty in our country, we need to make a number of assumptions about our current situation in the United States in relation to what we can and cannot do to get all Americans out of poverty.

First, the economic system of capitalism will not provide enough good-paying jobs for all of the adult people in our country who are capable of working and want good-paying jobs. Recently, the unemployment rate in our country stood at 6.4% ("Unemployment Hits," 2003). If we rely only on capitalism to solve our poverty problem, we will not solve it due to cycles of moderate to high unemployment. Consequently, if we want to solve the problem of poverty, we will need to do something more than just rely on capitalism.

Second, even when capitalism produces many new jobs and the unemployment rate goes down to 3% overall, which is roughly as low as it typically goes, we cannot assume that the jobs that are created pay above the poverty line. For example, Sanders (2000) noted that "30 percent of American workers earn poverty or near-poverty wages" (p. 3). This raises the question of how to get people out of their poverty in an economic system where 30% of the jobs pay at or below the poverty line. Because capitalism produces a considerable proportion of jobs below the poverty line, we will need to do something other than just rely on capitalism.

Third, even during the best of times, because our economy does not produce either enough jobs in total or enough jobs that pay above the poverty line, the government—local, state, and federal—will need to be a part of solving poverty in at least three ways: (a) by providing subsidies that bring the poor above the poverty line; (b) by providing social services, such as health care and retirement benefits, that the poor are not getting from jobs below the poverty line; and (c) by collecting enough tax money from those of us who are not poor to provide for these subsidies and services.

Fourth, there are probably some lower income people who are poor because they are lazy. Our country will need to face the problem of how to deal with these people. Even though these people represent only a small number of the overall number of poor people, we will need to address this problem.

Fifth, because there are different groups that make up poor people (e.g., people without jobs; people with jobs that pay below the poverty line; people who cannot work such as children,[1] the elderly, and the disabled; people who do not want to work and do not work), we will not be able to solve poverty by creating just one solution such as creating more jobs. Instead, we will need to create a number of solutions to get all people above the poverty line.

Sixth, to solve the problem of poverty in the United States, we will need to solve it within the context of a capitalistic economy. Some Americans may want us to go from capitalism to some form of socialist economy where all adults are guaranteed jobs. As a result, there would be no unemployment. I do not think that this is a realistic option now or in the near future. The simple reason is that so many of us who are not poor have benefited from, and are still benefiting from, capitalism with nice homes, good incomes, and many consumer goods and services. Too many of us currently benefit from the existing economy to realistically consider changing to socialism. Furthermore, from what we have heard and read about the standard of living in the Soviet Union during the Soviet era (1917–1990) and what we have read about the East European countries under the control of the Soviet Union and what has occurred so far in China and Cuba, the people living in these countries have had standards of living considerably below those of us living in capitalistic countries and have also had less freedom. Consequently, given our hesitancy to change to a socialistic economy, we will need to consider, for the time being, solutions that we can carry out within a capitalistic economy.

Seventh, given that capitalism cannot solve poverty by itself, and given the need to add the help of the government, we, as taxpayers, will need to give some of our resources in the form of taxes to solve poverty. This is not a pleasant situation to face, but I see no other realistic way given what I have stated in the preceding assumptions. As to who pays more and how much, this will need to be decided by Congress, state legislatures, and city and county governments. It would seem, however, that the most humane, fairest, and least hurtful way of collecting the needed taxes would be to tax people less if they are nearer to the poverty line and tax people more if they are further from the poverty line. In other words, if we want to get all people above the poverty line, we will need to allocate more tax revenue to address poverty, and one way of doing this is to have a more progressive income tax.

Eighth, as we solve the problem of poverty with the input of both capitalism and government, we should also see a decline in the severity of other social problems that are influenced by poverty such as selling drugs to make money to survive in poverty, engaging in prostitution to make money to survive in poverty, burglarizing homes or businesses to make money to survive in poverty, robbing people or businesses to make money to survive in poverty, and using a lot of drugs or drinking a lot of alcohol as a way of escaping the reality of one's poverty and hence being less capable of getting out of one's poverty. All of these problems are exacerbated by the existence of poverty in people's lives. Consequently, as we solve poverty in our country, we should also see a decline in other social problems.

Ninth, given that solving poverty will help to solve or ameliorate other social problems, it may be prudent to focus on poverty as one of the first problems we address. That way, we would get more "bang" for our invested "buck." It is at least something to consider given that we have many social problems but only a limited number of resources to use.

What Can We Do?

What can we do? There are a number of actions we can take to solve the problem of poverty. In this section, we discuss a number of steps we can take that are realistic and humane and will not cause people to sacrifice a lot of their resources, although some of us who are not poor will need to sacrifice some of our resources for the benefit of lifting poor people out of poverty. As I suggest various policies, see whether you think that these policies will get our poor people out of poverty and be as little sacrifice as is possible on those of us who are not poor.

Note that these solutions will not give poor people a middle-class lifestyle, but they will help them to live a better lifestyle than they do now. The solutions that could get our American poor out of their poverty are the following: (a) raise the minimum wage; (b) increase the child care subsidy so that parents of children who want to work, and are physically able to work, can work; (c) increase the earned income tax credit to make up for the gap between the minimum wage and the poverty line; (d) decrease the income tax, social security tax, and sales tax on the poor, near poor, and working poor so that they have more take-home pay; (e) increase unemployment compensation, provide more and better job training, and create more effective job placement services given a capitalistic economy where people lose their jobs due to corporations moving to other countries, downsizing, or going out of business; (f) provide corporations with tax incentives to create new jobs in our country; (g) consider a guaranteed minimum income that will help to keep people above the poverty line when our economy does not produce enough jobs; (h) improve the housing subsidies for poor people so that they can live in adequate housing; (i) create more public transportation facilities so that poor people who cannot afford cars can get to where jobs are; (j) provide adequate health care for the poor, the near poor, and the working poor who do not have health care; (k) provide good public schools so that poor children have a much better chance of getting out of their poverty than they now have; (l) provide the elderly who are poor with a prescription drug subsidy so that they can spend their social security and retirement incomes on everyday expenses; and (m) provide the preceding combination of subsidies and services to help those who are not capable of having jobs, such as the elderly and the severely handicapped, to support themselves and live above the poverty line.

Given these social policies that our country could carry out, we could go a long way toward solving the problem of poverty and also toward addressing other social problems related to poverty. Let us now elaborate on each social policy in turn.

Raise the Minimum Wage

Can we solve the poverty problem by providing poor people with minimum wage jobs? If a mother of two children gets a job at minimum wage ($5.15 per hour) and works 40 hours per week for 52 weeks per year, she will earn $206 per week or $10,712 per year. The problem is that this yearly income does not raise her and her children above the poverty line, which is currently $15,927 for a family of three (U.S. Department of Health and Human Services, n.d.). This mother is still below the poverty line by $5,215.

What can we do? If we raise the minimum wage to the poverty line for a family of three (i.e., $15,927), this would mean that a mother would need to have a wage of $7.66 per hour. So, this would be an increase of $2.51 per hour. This new minimum wage would be very helpful for mothers throughout the United States. However, that would mean a substantial increase in labor costs for employers. To pay this new wage, the employer will not be able to hire as many workers and/or will need to take a cut in profit. Employers do not want to make less profit, which could mean lower salaries for them and lower dividends for shareholders.

So, what can we do? Do we keep the minimum wage as it is and hence keep people thousands of dollars below the poverty line and yet we provide more minimum wage jobs and provide more profit for employers and shareholders? Or do we raise the minimum wage so that poor people can get out of poverty but we provide fewer minimum wage jobs and provide less profit for employers and shareholders? Here we see, as we discussed in Chapter 2 regarding the barriers to solving our social problems, the problem of vested interests of different groups and how these vested interests can help or hinder certain social policy intended to solve a social problem.

Another argument against raising the minimum wage is that it is inflationary. Yet on the other hand, many nonpoor people in their jobs expect to get certain percentage raises each year, and this also causes inflation. So, this raises the following question: Is it okay for the rest of us to get pay raises, even though they are inflationary, but not okay for poor people making a minimum wage to get a pay raise, even though it too is inflationary? Maybe a realistic compromise might be to increase the minimum wage somewhat (say, by at least $1.00) so that the new rate is not so hard on employers and does not cause a large decrease in the number of minimum wage jobs and yet is still an increase so that those workers who are making the minimum wage will at least get closer to the poverty line even though they will not reach it.

Another step that we could take vis-à-vis the minimum wage is to tie it to the yearly rate of inflation. That is, as inflation increases, say 3% per year, the minimum wage will also go up 3%. Currently, Congress needs to pass a law every time it wants to increase the minimum wage. What typically happens is that Congress gets pressure from business interests not to increase the minimum wage because that would increase their costs and decrease their profits. As a result, the minimum wage typically does not increase for years.

Yet inflation continues to occur each year. Over the years, people making the minimum wage are making less and less money because yearly inflation eats away at the value of their wage. The prices for food, rent, gasoline, shelter, clothing, electricity and water bills, insurance on homes and cars, and so on continue to go up each year, but their wage stays the same over 5 to 8 years or more; for example, the minimum wage has not gone up since 1997 (U.S. Department of Labor, n.d. a). If, instead, the minimum wage were tied to yearly inflation, poor people working for the minimum wage would not lose the value of their income each year. You are probably already thinking, "There is no way that owners of businesses will ever want this to happen, so they will vigorously lobby Congress to stop it from happening." I agree with you. That is the reason why we have not seen the minimum wage tied to the yearly rate of inflation. So, yes, this would be a tough sell to Congress, but it would be an option for getting people out of their poverty.

Increase the Child Care Subsidy

This raises the following question: How do we get a mother and her children out of poverty when she is already working full-time? Moreover, this mother will, in all likelihood, need to pay someone to take care of her children while she is at work. Typically, this will cost a minimum of $50 to $100 per child per week. This additional cost for child care with a minimum wage job will decrease the weekly income she brings home; for example, she will earn a minimum wage of $206 per week if she works 40 hours per week minus $100 in day care for two children, leaving her with $106. As you know, trying to keep together a family of three with $106 per week to pay for rent, food, transportation, and many other expenses will be extremely difficult—if not impossible. Rent alone can eat up what she has left. It is no wonder that families living at this level of subsistence face many problems that the rest of us do not face.

Our state governments and the federal government have recognized the problem of the cost of child care for low-income parents and have provided funds to help pay a portion of the child care costs. This help with mothers' child care expense gives mothers on welfare an added incentive to go to work, even with a minimum wage job. The problem, however, is that with the downturn in the economy during recent years, state governments have needed to hold the line on what they can afford to give out in child care subsidies. For example, the state of Kentucky has a waiting list of 2,700 parents who have applied for child care funding, but the state is unable to satisfy their requests.[2] This lack of funding makes it harder for these parents to get off of welfare and hold a job or, if they currently have a minimum wage job, to continue to work without going back on welfare.

This raises the following question: Do we provide more funding for child care for mothers who have minimum wage jobs? This question raises a difficult dilemma for all of us in our country because our federal government is currently spending more than it receives in taxes and therefore getting further into debt and because President George W. Bush got a tax cut through Congress, meaning that we will, at least in the near future, have even less tax money to pay for things such as child care subsidies. Do we increase child care subsidies even though it means going deeper into national debt? Do we maintain current spending for child care subsidies and leave a lot of low-income mothers on welfare and in poverty? Do we decrease child care funding even though we know that the consequences will be that more mothers will need to give up their jobs and go back on welfare? If more mothers go back on welfare, you and I will pay our tax money for welfare instead of child care funding. Do we have mothers return to welfare even though our policy since 1996 has been to get people off welfare and on to work?

If we want to solve the problem of poverty in our country, it seems that we will need to continue to provide, and even increase, child care subsidies to include more mothers who want to get off of welfare and out of poverty. Helping poor people who have minimum wage or near minimum wage jobs with their child care expenses seems to be another key element in helping poor Americans to get out of their poverty.

Increase the Earned Income Tax Credit

In addition to increasing her minimum wage, is there any way that a mother with two children can have enough money for her family to be close to, at, or above the poverty line? Yes, there is a way. When she files her income tax return, she can apply for what is known as an earned income tax credit (TaxCreditResources.org, n.d. b). This is money that the federal government and some states[3] give to people who work but still have below poverty-level incomes. The purpose of the earned income tax credit is to help get poor working people closer to the poverty line and give them "added incentives to work" (TaxCreditResources.org, n.d. b).

The earned income tax credit was created in 1975 by Congress but was expanded in 1986, 1990, and 1993. It expanded during the 1980s and 1990s under two Republican presidents and one Democratic president: Ronald Reagan, George Bush (senior), and Bill Clinton (Wilson, 1997, p. 222). One source asserted, "EITC [earned income tax credit] lifts more working families above the poverty line than any other federal program."[4] Wilson (1997) noted that these increases in the earned income tax credit "reflected a recognition that wages for low-paying work have eroded and that other policies to aid the working poor—for example, the minimum wage—have become weaker" (p. 222). Even with increases in the earned income tax credit during the mid-1990s, it still fell short of making up for "the sharp drop in the value of the minimum wage and the marked reductions in AFDC [Aid to Families with Dependent Children (welfare)] benefits to low-income working families since the early 1970s" (p. 222). Even when food stamps were added, workers' incomes still fell below the poverty line during the mid- to late 1990s (p. 222).

If we want to solve the poverty problem in the United States, we could make sure that adults who have minimum wage jobs will receive an earned income tax credit that will put them above the poverty line. As we discussed in the preceding subsection, we could also tie the minimum wage to the rate of inflation and then make the subsequent adjustment to the earned income tax credit so that all working people could rise above the poverty line, thereby getting many Americans out of poverty.

Decrease the Income Tax, Social Security Tax, and Sales Tax on the Poor

We can give poor people more take-home pay each week by decreasing the taxes that they pay. One way to decrease taxes on the poor is to make our income tax, social security tax, and sales tax more progressive, where poor people pay less in taxes and rich people pay more in taxes. Let us take the social security tax as an example. Currently, Americans who make $94,200 or less pay 6.2% of their income in tax. People who make more than $94,200 in income do not pay any social security tax on the amount over $94,200. We could change this tax to make it a progressive tax where, for example, the poor pay 1% in social security tax and the rich pay 8%. By the poor paying only 1% instead of 6.2%, they would increase their take-home pay by 5.2%, thereby helping them come closer to getting out of their poverty.

We could also decrease the state sales tax for poor people. For example, the sales tax in Indiana is 6%. A poor family of four needs the same amount of toothpaste, toilet paper, shampoo, and soap as does a rich family of four. When these two families go to the store to purchase these items, they both pay the same amount of sales tax. This sales tax takes a larger proportion of money out of the income of the poor family than it does out of the income of the rich family. We could pay a sales tax dependent on the income we make. That is, the less income we earn, the less sales tax we pay; the more income we earn, the more sales tax we pay. This would make the sales tax progressive and, in so doing, help poor people to get out of their poverty.[5]

We currently have somewhat of a progressive tax in place with respect to the income tax we pay. We could make the income tax much more progressive, where poor people would pay a lower percentage of their incomes in tax and rich people would pay a higher percentage of their incomes in tax. Taken together, if we made income tax, social security tax, and sales tax more progressive, we could provide more money for people to help them get out of their poverty.

Continue to Improve Unemployment Compensation, Provide Better Job Training, and Create More Effective Job Placement Services

Whenever we have a capitalistic economy, a number of people will lose their jobs through no fault of their own; for example, corporations move elsewhere, add robots to the assembly line, and downsize. We live in a society and world where corporations will put their own survival and profits ahead of the welfare of their workers. As a consequence, some people will lose their jobs and fall into poverty.

Unemployment Compensation

To get people out of poverty, we need to provide them with services as soon as they lose their jobs. One step we could take is to provide enough money for them to pay their daily, weekly, and monthly bills. Currently, we have unemployment compensation for people when they lose their jobs. This program was established in 1935 with the passage of the Social Security Act, where workers who lose their jobs "through no fault of their own" (Cornell Law School, n.d.) receive a percentage of their individual earnings within 2 weeks of filing for benefits. These benefits can last up to 26 weeks. However, during periods of high unemployment when workers have used up these 26 weeks, extended benefits can go into effect for an additional 13 weeks.[6] The weekly benefit that the unemployed worker is supposed to receive is between 50% and 70% of the worker's previous wage. However, data show that in actuality "the national average weekly benefit amount as a percent of the average weekly covered wage was only 35 percent" (Almanac of Policy Issues, n.d.).

This raises the following question: Can people live on 35% of their former income and receive $215 per week, on the average, and pay for food, rent or mortgage, and electricity, water, and other weekly and monthly bills? We need to review this amount to see whether it provides enough for people to live on until they can find another job.

Job Training

We also can get people quickly connected to retraining schools to help them learn new skills that are in demand in the current economy. As we all know, a capitalistic economy is constantly changing, with some jobs decreasing in demand and other jobs increasing in demand. Our job training programs need to constantly readjust as our economy changes.

Currently, people can take different paths in getting retrained. They can go to vocational schools that teach specific skills, they can go to employers who will train them, or they can go to a college or university to get a broader education in the form of a bachelor's degree.

Local administrators of these job training programs can make recommendations as to how we can have more effective training programs. We can also make these services more visible and more known through television, radio, and newspaper advertising that tells poor and unemployed people how and where they can get job training. Moreover, our schools can do a better job of teaching high school students how to write a résumé, how to search for a job, how to apply for a job, and how to interview for a job (Almanac of Policy Issues, n.d.). That is, high schools, in cooperation with each state's job training and placement services, could help graduating students to find the jobs they want, thereby lessening the amount of time they would be unemployed.

Job Placement

As for job placement services, we currently have in our country various kinds of services that help various kinds of people get jobs. For example, in Indiana, Transition Resources helps migrant and seasonal workers get connected to jobs (Almanac of Policy Issues, n.d.). Older workers who are looking for jobs can get help from a service called Experienced Workers (Almanac of Policy Issues, n.d.). There are also job matching services that match people to local jobs or to jobs that are available statewide. More recently, a national job bank called America's Job Bank helps to give people access to jobs at the national level (Almanac of Policy Issues, n.d.).

So, to get people out of poverty more quickly, we could increase the amount of unemployment compensation while they are unemployed, we could work to get unemployed people into job training programs and make sure that these training programs are changing along with changes in the economy, and we could work to improve local, state, and national job placement services. The improvements that we could make in these three areas could help newly unemployed people who have fallen into poverty, or other people who have been in poverty for a longer period of time, to get out of their poverty more quickly. This quicker turnaround time from unemployment to employment will also mean that the rest of us will pay less in taxes that go toward providing for unemployment compensation.

Give Corporations Tax Incentives to Create New Jobs

Even though people are trained and are looking for jobs via various job placement services, we need to find ways to create more jobs for the many people in our country who are unemployed. One important way to do this is to give corporations incentives to create jobs in our country. Corporations are in business to make profits. If they do not make profits, they will be out of business rather quickly. So, if corporate managers think that creating more jobs in our country will increase their profits, they will do so.

One way to stimulate job creation by corporations is to decrease the taxes on corporations and to make up for this tax loss by making the income tax more progressive by having the wealthy pay more in taxes and the poor and working classes pay less in taxes. Local, state, and federal governments could give more tax breaks to corporations in exchange for corporations providing a certain number of jobs in the United States.

Provide a Guaranteed Minimum Income

If the economy cannot provide enough jobs for all Americans who want them, we might need to consider giving some kind of income to people so that they can survive and stay above the poverty line. Capitalism has cycles of more or less unemployment. During the Depression of the 1930s, the unemployment rate rose to more than 20%. During the boom years of the mid- to late 1990s, unemployment was at times less than 4% (University of Texas, n.d.). Starting in 2000, our economy began to level off and take a nosedive. By July 2003, the unemployment rate had reached 6.4%.

What do we do when people cannot get jobs? Gans (1995) discussed the economy, jobs, and poverty in *The War Against the Poor* and asserted that we might need to face the fact that our economy will not produce enough decent-paying jobs for all who want or need them. He stated that due to the loss of factory jobs to other countries, due to the use of robots and computers in place of human workers, and due to corporate downsizing of midlevel jobs, we face a big challenge of producing enough jobs for everyone (p. 133). As a result, we might need to have other alternatives in mind if we wish to lift the poor out of their poverty.

If, in such a scenario, the government did not respond with some type of safety net of both income and services, poor people could not get out of their poverty and could feel a sense of despair and anger, especially in our country where we emphasize upward mobility. Moreover, the poor could question the legitimacy of the prevailing social system, form various kinds of conflict groups, and promote conflict in some way (this is a good example of where our theory of conflict and social change and the accompanying causal model outlined in Chapter 1 can be applied).[7] Thus, the stability of the country depends, in part, on matching the ideology of upward mobility with actual chances for people to be upwardly mobile. If jobs are not forthcoming, we might need to rely on the government to provide some sort of safety net of income and services.

One safety net that we could provide is a guaranteed minimum income that puts people above the poverty line when, through no fault of their own, they cannot find jobs and yet need money to pay their bills. Possibly, a guaranteed minimum income could begin at the point where people have tried to get jobs and their unemployment compensation has run out.

We could construct the guaranteed minimum income in such a way that it is more in people's vested interests to find work than to remain on the guaranteed minimum income. In other words, when they find work, they make more money at their new jobs than they would if they remained on the guaranteed minimum income. This could be one of the final methods that we use to get people out of poverty if the other methods we discussed do not totally solve the problem of poverty.

The guaranteed minimum income has been suggested by conservatives as well as liberals. At first thought, you might think that this is "just another liberal idea" where we, as taxpayers, need to pay more in taxes to provide for yet another government program. In reality, former President Richard Nixon, a Republican conservative, suggested that we implement the guaranteed minimum income. Also, the conservative economist Milton Friedman, in his book *Capitalism and Freedom,* supported this idea (Friedman, 1962, pp. 190–195). Friedman (1962) asserted that there are a number of advantages to having a guaranteed minimum income. First, it would be "directed specifically at the problem of poverty" (pp. 191–192). Second, it could act "as a substitute for the present rag bag of measures directed at the same end" (p. 192). Third, it "could be far less costly in money" (p. 193). Being a conservative and not liking big government and many government programs and the waste and overlap that these programs can bring, Friedman concluded that the guaranteed minimum income is the best way to address poverty. Finally, we might point out that Harold Kerbo, a specialist in the study of social stratification, noted, "The United States is the only industrial nation that does not have a basic guaranteed income program for all families below the poverty level" (Kerbo, 2006, p. 37).

If we implement the guaranteed minimum income, this is how it could work. We, the American people, would decide how much of our tax money we want to give to put poor people above the poverty level. Let us say, for example, that a mother of two children was laid off from her minimum wage job, where she was making $10,712 per year, but the poverty level for her was $15,927. She was on unemployment compensation, and her time on this has been exhausted. She could receive a guaranteed minimum income that puts her and her children above the poverty line, for example, $16,000. If she eventually finds a part-time job that gives her $1,000 more income, we would not take away the equivalent amount of $1,000 because she would have no incentive to work. To give her an incentive to work, when she earned $1,000, we would take, for example, only $500 away from the $16,000 guaranteed minimum income. As a result, she would get $15,500 of guaranteed minimum income and $1,000 from her job and now make $16,500. That is, she would make more money working than she would not working. If she was able to work more and made another $1,000, again only $500 would be subtracted from her guaranteed income, and she would now make $17,000. This same process would continue. As the mother works more and more, she makes more and more, thereby giving her more incentive to work than not to work.

If the economy cannot provide enough jobs for everyone, we might some day be in the position that Gans predicted. Should we, as a society, consider the guaranteed minimum income? This is an option to which we will need to give serious thought and discussion because it is one that could solve our poverty problem, especially if we find in the future that our economy provides relatively fewer and fewer jobs.

Improve Housing Subsidies

What should we do for people who live in dilapidated housing or have no housing? Even if we raise the incomes of poor people to a little above the poverty line, in many instances they will not have enough money to live in adequate housing.

Due to the Depression during the 1930s, when many people were out of work and had a difficult time in finding decent housing, the government began to build housing for people as a way not only to provide housing but also to provide jobs for people and therefore stimulate the economy. Slum housing was cleared and 114,000 low-rent units were built (Palen, 1997, p. 320). Although this program ended during World War II, it was fairly successful in that it built housing that poor people could afford.

With the Housing Act of 1934, which created the Federal Housing Administration (FHA) and the Veterans Administration (VA), both of which insured home loans, banks began to loan more money to people who were starting to improve economically and therefore began to move out of public housing. The people getting these loans began to move to new homes being built in the suburbs. The FHA and VA, however, strongly discouraged giving loans to African Americans so that people buying new homes in the suburbs would, they thought, have their property values protected. The results of these loans and discriminatory policies were as follows. First, white families who were improving their economic situations moved out of the public housing units to the suburbs. Second, black families had no option to borrow money and build new homes in the suburbs and thus had no chance to own new homes and begin to build equity in their homes and hence use the building of the new homes as a way to build wealth. Third, the people who were left in the public housing units were more likely to be less successful economically, meaning that poor people became more physically concentrated in public housing. Fourth, the children who remained in the public housing saw fewer and fewer successful role models who were making their way up and out of poverty through legitimate means versus through illegitimate means such as robbery, selling drugs, and prostitution. Fifth, the general public in the United States began to become more negative about public housing, with the very poorest of the poor making up the vast majority of residents of public housing, thereby giving public housing a negative stigma compared with public housing in Europe, where many working- and middle-class people lived. White families were also given a tax break on the interest they paid on their mortgages (home loans). The loans together with this tax break allowed white families to begin building wealth that was not available to African American families.

More recently, our country has provided vouchers to poor people so that they can go to various privately owned apartment buildings and get an apartment and pay 30% of their income toward rent while the federal government pays the landlord the rest of the rent. This program is called Section 8. The advantages of Section 8 are as follows. First, poor people can choose various places to live, thereby giving them some freedom of choice of apartments and geographic location. Second, given more freedom of choice of geographic location, poor people no longer need to be physically concentrated in one area and hence can have more choice of neighborhoods and schools. Third, poor people now have a chance to live closer to their jobs. Fourth, the federal government does not need to be in the business of building and running a housing program.

Currently, giving poor people vouchers so that they have choice and the ability to move to various places could be the best answer for helping poor people to attain adequate housing. However, because 5 million families still live in inadequate housing,[8] we could expand this program so that more poor people could have adequate housing. If we do not provide enough Section 8 funding for poor people, it seems that the only other alternative to provide enough adequate housing for poor people is for the government to get back in the business of building low-income housing. The problems with this are that the government adds more bureaucracy in building and maintaining such housing and that the poor end up living in areas of the government's choosing where there are higher concentrations of poor people and fewer chances for employment. Given these problems, the Section 8 program seems to be a better idea.

Given President Bush's tax cuts, it will be more difficult to find the tax money to expand the Section 8 housing program. Time will tell what we will do with respect to housing for poor people. In the meantime, millions of poor Americans will live in very poor housing or have no housing.[9] It is up to us to decide what to do. What do you think we should do?

Create More Public Transportation

As I mentioned in the preceding subsection with respect to providing people with Section 8 vouchers so that they can have choices as to where to live, these vouchers will also help the transportation problems that many poor have because they cannot afford to own a car.

So, how do people in large and medium-sized cities get to their jobs, grocery stores, and other places they need to go (e.g., doctor appointments, dentist appointments)? If they cannot afford a car, what do they do? Ideally, they live close enough to walk to jobs, grocery stores, schools, hospitals, and banks. But in the spread-out fashion of the contemporary city, where services many times are miles apart, transportation for poor people becomes a major challenge. Public transportation in the larger cities helps to solve this problem for many poor people.

One solution for poor people in larger cities is that the public transportation system could be expanded. For example, during recent years the city of San Francisco has created a subway and train system and has continued to expand both systems to take care of the transportation needs of more and more people and, at the same time, allow more people from one part of the Bay Area get to jobs and services in other parts of the Bay Area.

The extending of these transportation systems should be of great help to poor people in getting to jobs and services.

Besides helping poor people to have better chances at getting to new jobs, these public transportation systems allow many nonpoor people to use this service and hence cut down on the use of cars that contribute so much to the air pollution in our country. So, this partial solution to poverty will also be a partial solution to our environmental problem (see Chapter 13 for a discussion of our environmental problems).

What can we do in smaller cities and rural areas? One program that has been implemented in southeastern Indiana, called "Catch-a-Ride" and covering a five-county area, offers transportation for poor people who need to get to work (personal interview with M. Hueseman, Catch-a-Ride, Lifetime Resources, August 31, 2003). Based on the Catch-a-Ride model, we could design a national program that covers smaller cities and rural areas to help the poor travel to their places of work.

Provide Adequate Health Care for the Poor, Near Poor, and Working Poor

During the 21st century, even after much discussion by President Clinton during the early to mid-1990s about the need to provide all Americans with some form of adequate health care, our country had 44.3 million people who did not have health insurance (Pear, 1999). Many Americans are in a crisis situation if any of their family members gets seriously ill or has a serious accident, where the illness or accident will cost the family thousands of dollars. Most of these families without health insurance are poor, near poor, or working poor, where the jobs the parents have do not provide health insurance benefits. Because these people are poor or close to being poor, they cannot afford to buy health insurance on their own. So, they take a chance and go without it.

We are the only industrialized country that does not have a "national health program meeting the medical needs of all families" (Kerbo, 2006, p. 37). Providing health care for all Americans can be a key ingredient in helping poor people to get out of their poverty, thereby addressing three social problems discussed in this book: rising inequality, persistent poverty, and no health care for millions of Americans.

Create Good Public Schools

To give poor children a chance to get out of their poverty, we could make sure that we create good public schools for all students. By *good schools*, I mean a number of things. First, I mean that all children are in a safe environment when they are at school. Second, I mean that all public schools are well maintained and have up-to-date facilities and equipment so that students get a very good education (and are also proud of their schools).

Third, I also mean that all of the teachers are well qualified by being certified in the areas they teach. Moreover, these schools have low student-to-teacher ratios so that teachers can take more time with each student. For example, schools in poverty neighborhoods could have student-to-teacher ratios of 10 to 1, where teachers can concentrate on bringing each child up to his or her full potential.

In such an atmosphere, children from poor families will get a good education, which will, in time, help them to go on to college or trade school or prepare them to work right after high school. To provide schools that are safe and well maintained, have up-to-date equipment, have well-qualified teachers, and have low student-to-teacher ratios will require that we invest more money into our public school system.

If we want to work at getting rid of poverty, we will need to create a much more proficient public school system. Creating such a system will be a major way to help poor children acquire the proper education and skills to get out of poverty and make it on their own, thereby addressing two social problems in our country: poverty and poor-quality schools (see Chapter 7 on unequal education in our country).

Give the Elderly a Prescription Drug Subsidy

A group of poor people who will not be helped with an increase in the minimum wage, a larger child care subsidy, or better quality schools are the elderly poor. We could create other ways to get them out of poverty so that they can live a decent lifestyle.

A number of things could be done to help the elderly poor get out of their poverty. Many of the elderly have social security income that helps but might not keep them out of poverty. Medicare, a health program provided by the federal government, is a great help in providing the elderly with health care. Yet currently the prescription drugs that the elderly must pay for take a substantial chunk out of their monthly incomes. Congress has been struggling with this issue. It appears that Congress has arrived at a partial solution where the elderly will get some help from the government, and this will take some financial pressure off of the elderly (Pear, 2003). However, more could be done in the prescription drug area to help the elderly poor rise out of their poverty. Moreover, a national health care system (discussed in Chapter 10) could address our health care problem and, at the same time, help the elderly poor to get out of their poverty.

Concluding on an Optimistic Note

If we want to do so, we can decrease poverty in our country considerably. The reason why I say this is that Western European countries have already done this. Whereas the poverty rates in countries such as Germany, France, Belgium, Denmark, The Netherlands, and Sweden are roughly the same as our poverty rate (as measured by income below 50% of median income in a nation), *all* of these countries' poverty rates declined to 5% to 7% levels *after* they added income and services provided by the government, whereas our country's poverty rate remained much higher at 19% (Kerbo, 2006, pp. 260–262). If European countries can do this, we can do this too. This brings us to the following question: Are we really willing to get rid of poverty in our country? We have the tools discussed in this chapter to do this.

Questions for Discussion

1. Should we increase the minimum wage?

2. Should we increase the child care subsidy?

3. Should we increase the earned income tax credit to the poverty line?

4. Should we tax the poor less in terms of income tax, social security tax, and sales tax while we tax the rich more?

5. Should we tax corporations less to give them incentives to stay in the United States and provide more jobs for American workers?

6. What should we do with respect to unemployment compensation, job training, and job placement?

7. Should we have a guaranteed minimum income if our economy cannot provide enough jobs for all Americans who want jobs?

8. What combination of the previously discussed measures do you think will eliminate poverty?

9. Which of the solutions discussed in this chapter will be tried and which will not? Why?

10. If you were asked to eliminate poverty, what would you do?

5

How Can We Solve the Problem of Racial/Ethnic Inequality?

The year was 1954. I had just turned 10 years old. My aunt, uncle, and cousin had been to Miami Beach, Florida, the summer before and were excited about vacationing in Florida again and wanted our family (my mother, father, older brother, and me) to go this summer. My father was not sure that he could afford the trip (recall that at that time many families adhered to the traditional gender roles where the father worked at a job and the mother stayed home to work at cooking, cleaning, doing the laundry, grocery shopping, and doing the endless ironing that I remember seeing my mother do in the kitchen). With just one income, and that income being a high school teacher's income, we were not poor but we did not have much left over after paying the monthly bills. My father finally decided that with the money he earned being the secretary of the Kiwanis Club and taking wedding pictures during the summer (he was a wedding photographer during the summer to make ends meet), and saving a little bit here and there, we could go. To say the least, as a 10-year-old boy who had never been out of Indiana, I was overjoyed! On the drive down to Florida, I would get to see a big river called the Ohio River, I would get to see what mountains really looked like when we went through the Smoky Mountains, and I would get to see my first Civil War battle site on top of Lookout Mountain in Tennessee. When we got to our destination, I would get to stay in a motel (motels were something new at that time), play on the beach, dive into the waves, and swim in the ocean.

I was so excited to see and experience all of these things. What I did not realize at the time, nor did anyone else in my family, was that I was to have an experience that would leave an indelible mark on me for the rest of my life. We were on our way in the family car (very few people went on vacations by planes in those days), and we had already driven over the Ohio River into Kentucky and had seen the Smoky Mountains and the Civil War battle site. We were now in the middle of Georgia in peach tree country, where on both sides of the road were endless fields of peach trees and you could smell the aroma of peaches in the air. This is where, unbeknownst to me at the time, I had my life-changing experience.

We stopped along the roadside (nearly all roads had just two lanes at that time because there was no such thing as the interstate system) to get some gasoline at a small gas station and use the restroom. I got out of the car and went toward the restroom. The signs said, "White Men" and "White Women." There was a third sign that said, "Colored" and an accompanying arrow that pointed to behind the station. I went to the "White Men" restroom (I am white). On leaving the restroom, my curiosity got the better of me because I wanted to see what the "Colored" restroom was like. I walked behind the station and discovered that there was no restroom—just a field.

I was shocked. I ran to my mother and father, who were near the car, and exclaimed in an upset way, "The colored people do not have a restroom! Why is that?" My parents, I am sure, were somewhat taken aback and really did not know what to say. I recall my mother saying, "Honey, let's get in the car and we'll talk about this later." A few minutes later, because I was so upset that I felt people were not being treated equally and fairly, my parents tried to change the subject to calm me down. I eventually calmed down and resumed thinking about the beach, the ocean, the waves, and so on.

However, I never forgot that moment. I would not fully understand that moment and what was to transpire in my life until years later as an adult. Let me explain to you, using our sociological perspective, what happened in my life since then.

My parents were loving and caring and were considerate and respectful of all people. My father, unlike many other white people at the time, spoke to, knew, and talked with black people in my hometown and always did so with a spirit of dignity and respect. When he was away from them, he never talked down or against them. I grew up in this atmosphere of thinking, "That is the way you treat everyone—with dignity and respect." Recall in Chapter 1 the theoretical idea of differential association. My association was with my parents, who taught me to treat all people one way. Others, as I was to later find out, were taught to treat some people, especially those of another skin color, another way. One sociological insight here is that differential association can be, and usually is, much more influential on us than we realize. This idea also hearkens back to the idea by the great French sociologist, Emile Durkheim, who said, as you may recall from Chapter 1, that society is external to us and yet coercive on us. That is, as you and I are born, society is outside of us in the form of norms, values, laws, beliefs, and customs. All of these socially constructed things are eventually taught into us and hence begin to influence how we think, believe, feel, and act. Most of the time, we are not conscious of this process, but it happens to all of us. We call this process *socialization*. Typically, parents have the most influence on us via socialization, especially when we are young. Others, such as teachers, coaches, and religious leaders, will later have a role in socializing us and therefore have increasing influence on us as we grow older.

What happened to me that day was that, sociologically speaking, I began to be conscious of racial inequality for the first time in my life. In an abrupt and unexpected way, I learned about the existence of one kind of inequality, namely racial inequality. I saw it with my own eyes. No one could tell me differently. I was thinking and realizing at the time that the colored people were treated differently and unequally from the white people. In that moment, two things happened: I became conscious of racial inequality, and I became concerned about it. I became concerned because what I saw went against everything that I had been taught—at home, in school, and in my church.

I was to have a number of experiences thereafter that only reinforced what I had seen and unexpectedly experienced on that day decades ago. Over the years, I was to learn that, even in my hometown in Indiana, blacks lived in only two neighborhoods of the city. They never went swimming at the public swimming pool, and they were not in my church but rather had their own churches. I never saw them in a restaurant. They were not members of the Kiwanis Club, the Lions Club, or country clubs. They did not stay in the motel down in Florida, walk along the beach, or swim in the ocean, but they did work at the motel maintaining the air-conditioning, trimming the shrubs, and so on.

As I matured, I began to see and realize what these things meant. I observed many instances of segregation, meaning that the minority population is separated from the majority population in many different ways by the informal norms, customs, and laws at the time. At that time, I did not realize it, but I was beginning to look at life in a sociological way. I was not aware of the discipline called *sociology* that focused on the study of norms, values, beliefs, laws, and much more.

During my sophomore year in college, I took an introductory sociology course as a way to fulfill a requirement for the bachelor's degree. Again, I did not realize what would soon happen to me. The professor began to say, "We are going to study the society for the next 16 weeks, and among the things we are going to study are poverty, racial prejudice and discrimination, and … " I was immediately hooked. Those were topics that I had wanted to know more about all of my life—or at least since I was 10 years old when I had the experience at the gas station in Georgia. No other course had ever taught me about these subjects. I had been curious about these issues for years but did not know that there was an entire discipline devoted to the study of them. I was curious and had an insatiable desire to know as much as I could about phenomena such as racial prejudice and discrimination—what caused these things to happen, why they exist in a society where we are supposed to love one another, what all the consequences of prejudice and discrimination are, what will happen in the future, and what can be done.

Not too long after this, after taking a few more courses in sociology, I realized that this was what I wanted to do: study sociology, teach it, and share it with others. So, in a rather circuitous route that spanned a number of years, I had found my calling. In looking back, I became conscious and concerned on a summer day in Georgia years before. Later, I was to realize how curious I was about the study of social phenomena in general, and of racial inequality in particular, and also to realize that I had found my calling. This is how I got into sociology.

I use this personal story with the use of sociological concepts and theoretical ideas as a lead-in to address how we can solve the social problem of racial/ethnic inequality. Let us now discuss where we currently are in our society and where we could go.

Now and Where to Next?

Recall our theory of conflict and social change and the theoretical propositions and causal model in Chapter 1. During the mid-1900s, African American people became more conscious of their racial inequality. They started to communicate with each other and were soon led by a charismatic leader by the name of Martin Luther King, Jr. They formed a conflict group and began to have peaceful protests in the form of marches, demonstrations, boycotts, and sit-ins. They put pressure on presidents and members of Congress to pass civil rights laws that would allow all minority groups to use any public accommodations (e.g., restaurants, parks, motels), vote, hold public office, work at any job for which they were qualified, and live in any house and neighborhood they could afford. They sought to be able to join private clubs in which they were interested (e.g., country clubs, tennis clubs, Kiwanis Club, Lions Club, Rotary Club). They wanted to be able to travel anywhere in the United States and visit theme parks (e.g., Disneyland, state parks, national parks) and receive the same food and lodging accommodations that whites received. They wanted to have the same chance as whites to play college and professional sports and therefore to earn college scholarships and pro salaries.

African Americans did not make much progress in their attempt to gain racial equality until, as the theory of conflict and social change shows, they became conscious of their situation, saw the unfairness of their situation, questioned the legitimacy of their situation, formed conflict groups, and carried out peaceful conflict in the form of what is now known as the civil rights movement. Over time, as the theory predicts, African Americans were able to bring about a new social construction of reality. A new U.S. Supreme Court decision (*Brown v. Board of Education*) and the desegregation of buses, city halls, county courthouses, colleges and universities, sports at colleges and universities, various places of employment, and many kinds of voluntary organizations together created a new social structure in our country where, as the theory predicts, there was less prejudice and discrimination; less inequality of money, power, and prestige between minorities and the majority population; and a more humane and just society.

There has been substantial structural change since the 1950s in the form of new informal norms, including how we treat each other in everyday interactions, new desegregation laws (e.g., desegregation of schools, offices, motels, hotels, restaurants, parks, and the military), laws to allow voting for minorities, laws to allow for the holding of public office, and laws to provide minorities with opportunities to buy homes they can afford—new ways of applying the main values of our country, such as equal opportunity, justice, and freedom, to minorities and not just to the majority population of white Americans. So, the result has been substantial social change for African Americans in particular and for minorities in general because females, Hispanic Americans, Native Americans, homosexuals, the elderly, and the handicapped are other minorities whose members have benefited from structural change that came out of the civil rights movement by African Americans.

Where are we now? Although our country has made great strides in moving toward more equality for many different minority groups, we still have some distance to travel before we can say that minority Americans of all kinds have the same equality as do majority Americans. This raises the following question: Where can we go from here? That is, how can we move further in the direction of more racial/ethnic equality during the first half of the 21st century? In the coming pages, I make a number of suggestions that I believe can take us in that direction.

What Might Racial/Ethnic Equality Mean?

What might we mean by achieving racial/ethnic equality? Such equality could include some of the following changes. There is no longer racial/ethnic prejudice, meaning a negative attitude by one group of people that prejudges another group of people. There is no longer discrimination, meaning unequal treatment by the majority group of people toward a minority group of people. There is no longer institutional discrimination where the existing societal way of doing things works to the disadvantage of minorities. For example, we, as Americans, pay for our public schools through local property taxes, and this puts minority people at a disadvantage because they are more likely to live in poverty-laden areas of the country that cannot afford to collect as much tax revenue from property taxes as do whites and therefore cannot afford to have as good-quality public schools as do whites; therefore, racial/ethnic equality would mean that there is equal opportunity for minorities to attend good-quality public schools that prepare them just as much as white Americans to go to college or trade schools so that they are qualified to get decent-paying jobs. Attaining such racial/ethnic equality would also mean that there are comparable proportions of minorities in each social class as compared with the white population. Right now, there is a smaller proportion of African Americans, Hispanic Americans, and Native Americans in the middle, upper middle, and upper social classes and there is a larger proportion of minorities in the lower social classes. Attaining racial/ethnic equality would mean that minorities have similar proportions of their populations in various social classes as does the white majority.

Achieving all of these kinds of equalities together will indicate that we have reached racial/ethnic equality. Realistically, it may take a while to achieve all of these equalities, but these are goals toward which we can certainly work.

What More Can We Do?

I want to address two general areas that I believe we could address as a way to solve, or at least greatly decrease, racial/ethnic inequality. The first area deals with what we can do to decrease racial/ethnic inequality directly, whereas the second area deals with what we can do to decrease this type of inequality indirectly.

Direct Measures to Solve or Greatly Decrease Racial/Ethnic Inequality

Teach More Tolerance and Acceptance

Schools, Teachers, and Teaching. To address racial/ethnic prejudice and discrimination directly, we can increase the effort to teach tolerance and acceptance of other races and ethnic groups in schools. Teachers can emphasize the need for all Americans to go by some of the core values and beliefs of our society such as equal opportunity, fairness, justice, and freedom. Teachers can devise various teaching techniques to teach children that all Americans deserve to have these values and beliefs fully applied to them— not just to whites, middle-class people, or males. The more young students continually hear this message from kindergarten on into elementary, middle, and high school, the more such a consistent type of socialization is imprinted on the minds of young people, the more likely this type of socialization will be internalized and accepted, and the more likely any socialization of intolerance from the home environment or elsewhere can be negated.

Government. One way to address racial prejudice and discrimination is for the government to sponsor public service messages on television and radio and to place ads in newspapers, in magazines, and on billboards along highways. By taking such steps, the government can communicate clearly and visibly to the American people that such a stance represents the will of the American people. In other words, to hearken back to our theory of conflict and social change, the government establishes a new legitimacy; that is, it is seen as right to act in a tolerant and accepting way toward fellow Americans of another racial/ethnic group.

Private Organizations. Private organizations can sponsor various activities with the intent of teaching tolerance and acceptance. We already have numerous organizations that help various groups of people such as Big Brothers and Big Sisters, Habitat for Humanity, the Salvation Army, and Boys' and Girls' Clubs. Many of these organizations, in their own way, teach tolerance and acceptance now. These organizations, in cooperation with each other, can brainstorm to see what different ways they can work together to present a common, visible, and intentional theme of tolerance and acceptance. So, we have existing resources and organizations that could see what more they could do collectively in a local community to promote more racial/ethnic tolerance and acceptance.

We as Individuals. We, as individuals, in our daily lives can be constantly alert to opportunities that we personally have to say or do something that promotes racial/ethnic tolerance and acceptance. An easy step to take is the basic treating of every person we meet with dignity and respect. This is something that each person can do in his or her daily life. In other words, we can serve as daily role models of tolerance and acceptance.

Redress Grievances

We can continue to enforce the existing laws against racial/ethnic discrimination so that minority Americans know that they can go to court to have a racial/ethnic transgression redressed. This option can be made more well known by teaching about it in schools and by the government's incorporating such information into its public service messages.

Accept More Interracial and Interethnic Dating and Marriage

Up to and including the 1950s, there was a strong norm or taboo against whites and blacks dating and marrying. A few people did this, but most parents, friends, and others strongly advised against dating and marriage. Given such an atmosphere of informal norms and strong social pressure, most blacks and whites (and other combinations of minorities such as whites/Mexican Americans and whites/Native Americans) did not venture into dating. They might have seen and been around someone of another racial/ethnic group to whom they were attracted, but they knew their parents "would have a fit" if they even brought up the subject. So, for most people, the social structure of informal norms, customs, and family and community social pressure were, as Durkheim would put it, external to and yet coercive on anyone who might even consider asking out someone of another racial/ethnic group or even religion (during the 1950s, there was still a strong norm among many families that Catholics did not date or marry Protestants and vice versa, and "heaven forbid" dating Jews, atheists, or anyone else who did not have the "right" faith).

During the 21st century, we see more and more interracial/interethnic dating and marriage. In shopping malls, restaurants, movie theaters, and other public places, we see mixed couples. In schools, we see students who are of different shades and colors that represent mothers and fathers from different racial backgrounds. So, we are in the process of seeing and accepting more individuals and couples of mixed racial/ethnic and religious backgrounds.

As our society in general and as we, as individuals, get used to and accept more interracial/interethnic dating and marriage, this process will, over time, work to promote more racial/ethnic equality. We will get used to seeing and interacting with people of different combinations of races and ethnicities and will come to see this as less and less of a "big deal." Such interracial/interethnic dating and marriage and procreation of offspring will, I hypothesize, continue to break down barriers between people, and there will continue to be more tolerance and acceptance. To put it in Merton's (1967) terms, this process will act as a latent function for racial/ethnic minorities in that as there is more interracial/interethnic dating and marriage, minorities will be more accepted, will be more upwardly mobile, and will attain more money, power, and prestige. The consequence will be that they will increase their survival.

Promote a New Kind of Affirmative Action

The U.S. Supreme Court affirmed affirmative action in college admissions, to a degree, during the summer of 2003 (Greenhouse, 2003). In a 5 to 4 decision, the justices stated that race can be considered as one factor in admission to law school. The intent of the Supreme Court is that although admissions offices cannot use quotas with regard to race, they can use race as a factor in achieving diversity of a student body "because such policies promote cross-racial understanding and break down racial stereotypes" (p. A4) (stereotypes are generalizations about a group of people that are unfavorable and oversimplified).

Although affirmative action helps minority students to gain admission to undergraduate and graduate schools, helps to give them more opportunity to get ahead, and creates a more diverse student body that allows for more interracial/interethnic interaction and understanding, there are major criticisms of this method of attempting to solve racial/ethnic inequality.[1] First, it can leave out well-qualified white students who earn higher academic grades and SAT scores and have more extracurricular experiences than do minority applicants, thereby polarizing society with accusations of reverse discrimination against whites.[2] Second, it can help middle- and upper middle-class minorities who might not need the help. Rather, lower income minorities and whites are the ones who need help in having a chance to get ahead in our society.

Many people in our country have thought that we needed to go through a time of having affirmative action in college and graduate school admissions and in hiring for jobs because there was so much prejudice and discrimination within individuals and organizations (e.g., businesses, factories, unions, schools, churches).[3] In other words, we needed to take strong measures to break these barriers so that minorities could get a chance to go to college and get decent-paying jobs.

It seems that a number of African Americans have benefited from affirmative action and have joined the middle class, as indicated by their incomes and middle-class occupations. The children of these families will, like white middle- and upper middle-class families, be able to have sufficient resources during their childhoods so that they will be able to attend good schools, go to college and graduate school, and get good jobs without further assistance from affirmative action programs.

However, minorities who are still left out of the chance for upward mobility in our country, especially poor African Americans, Hispanic Americans, and Native Americans, are those who come from poor families and neighborhoods, regions, or reservations that have high poverty, high unemployment, and poor-quality public schools. So, problems continue to persist among poor minority Americans.

We can continue affirmative action, especially for poor minorities and whites.[4] This type of affirmative action would be more acceptable to most Americans because they would be more sympathetic toward poor people in general (e.g., black, white, brown, red, yellow) having the chance to get ahead—regardless of race/ethnicity.[5]

Get More Minorities Into the Middle Class or Higher Classes

The more we can get minorities into the middle class or higher classes, the less racial/ethnic prejudice and discrimination there will be because there will be more interaction and understanding between minorities and whites as well as acceptance of whites with minorities and of minorities with whites due to their living in the same neighborhoods, parents having similar kinds of occupations, parents joining similar voluntary associations, children going to the same schools, and children being involved together in sports, choir, band, and other extracurricular activities. All of these commonalities together will cause minorities and whites to have more things in common, and this should result in more acceptance of each other.

However, to get more minorities into middle or higher social classes, we can also carry out a number of indirect measures such as developing good-quality public schools for all Americans, creating more decent-paying jobs, building a tax system that takes less from poor and near-poor minorities and whites, and providing more social services such as public transportation and child care subsidies. All of these measures will create the social structural conditions for minorities to move into higher social classes, thereby creating more racial/ethnic equality.

Indirect Measures to Solve or Greatly Decrease Racial/Ethnic Inequality

Develop Good-Quality Public Schools

One indirect way we can create more racial/ethnic equality is to create good-quality public schools for all children in the United States (see Chapter 7 on unequal public education). Currently, we have unequal public education in our country, and this in turn perpetuates inequality.

The main reason for educational inequality is how schools are funded. Public schools in our country are funded mainly through people paying taxes on the property they own. These local property taxes pay for the building and maintenance of school buildings, teacher salaries, and books, computers, and other materials. The problem is that a disproportionate percentage of African American, Hispanic American, and Native American schoolchildren live in poverty areas where not as much property tax can be collected for each child. As a result, these children, along with poor white children, frequently do not have the quality of schools that middle- and upper middle-class children have. These young students are many times not academically prepared for trade school or college and therefore are not as able to get good-paying jobs and are more likely to end up in low-paying, minimum wage kinds of jobs that typically have few, if any, benefits such as health and retirement benefits.

In other words, to have the chance to be upwardly mobile in our society by having a good job, young people today, more than ever before, need to get a good education. Otherwise, they are more likely to remain in poverty. As you know, we have lost many good-paying unskilled factory jobs to other countries (called *deindustrialization*) because corporations can pay factory workers in these other countries lower wages with no health or retirement benefits and hence can make a larger profit. From the mid-1800s to the mid-1900s, our country had many decent-paying unskilled factory jobs that provided sufficient incomes for many American families. However, these days there are fewer of these kinds of jobs and more of the lower paying kinds of service jobs.

So, if we want to help poor African Americans, Hispanic Americans, and Native Americans—as well as poor whites—to have a chance at getting ahead and have a chance at getting a good job with a good salary, one major step our country can take is to create good-quality public schools so that poor minorities and poor whites have a chance to get ahead. By the way, it is also in our vested interests as a country to create such schools for all American children so that young people will be prepared to take the skilled kinds of jobs that our economy is creating such as those in the computer field, engineering, and medical technology.

Providing for such schools means that we, as a country, will need to invest more money in our public school systems from kindergarten through the 12th grade. With more money invested, we can hire more teachers and more qualified teachers, meaning teachers who are certified in the areas they teach (e.g., a biology teacher has a major in biology in college and has ample educational background to know how to teach biology). Many poor children of today, and poor minority children in particular, do not have certified teachers teaching them. Also, there might not be enough certified teachers for all public schools in our country because many people who go to college do not want to become teachers because they believe that teachers do not make much money. Even if there are certified teachers available, these teachers might not want to teach in schools that have overcrowded classrooms, do not have enough equipment, and/or are located in dangerous neighborhoods.

With more money, we can motivate more people to go into teaching and provide for enough teachers and enough certified teachers for all children. With enough money, we can create well-maintained school buildings for all children. With enough money, we can provide enough books, computers, and other materials for teachers to be able to do a good job and for students to be able to learn well and prepare themselves for college, trade schools, and graduate schools so that they can get better paying jobs with health and retirement benefits. Consequently, a major key to more racial/ethnic equality is to create excellent public schools for all children in the United States, and these in turn will provide minority children with a greater chance to be upwardly mobile and to be prepared to work at better paying jobs and attain a middle-class lifestyle.

This raises the following question: Where will the money come from? Currently, I see two possibilities that will help us to improve our public schools. One possibility is that we can make the federal income tax more progressive, especially on higher income people. Also, we could tax the wealth of rich people more. Either a more progressive income tax or taxing wealth more, or a combination of these ways, could supply the money needed for good-quality public schools.

The problem, as you might suspect, is that the wealthy will not want either their incomes or their wealth taxed more. Yet of all the places to get additional tax revenue to pay for schools, these two places would be the least hurtful to people.

Create More Decent-Paying Jobs

We need to find some way to have a strong and vibrant economy that provides enough good-paying jobs for all of the Americans who want them. If we do not create enough jobs or if we do not create enough good-paying jobs, Americans, even with a good education and well-developed skills, will be unable to get jobs appropriate for their educational and skill levels. So, this leads to the following crucial question: How do we create enough good-paying jobs in a capitalistic economy that does not necessarily provide enough jobs and enough decent-paying jobs?

One key method is to invest more money into what is called *research and development* with the intent of finding more ways to create good-paying jobs. That is, governments, corporations, colleges and universities, private foundations, and "think tanks" can invest money hiring people to do the following that can increase the number of good-paying jobs in our country: (a) invent new products to market and thus give new or existing businesses the reason to expand their plants and hire more people to produce these products, (b) invent new services that people would want to receive and thus create a new demand for jobs, (c) invent new ways to make a profit by recycling existing products, (d) fund research to see how our American companies can sell more goods and services to people in other countries, and (e) expand current services to meet the needs of people and, by so doing, create more jobs in the service sector; for example, expand the child care subsidy to a waiting list of 2,700 applicants who are working in low-income jobs in the state of Kentucky and, at the same time, add new jobs to administer this expanded service (Yetter, 2003).

A second way to create more decent-paying jobs is to have less tax on corporate earnings and more tax on personal income and wealth. The reason for doing this is that corporations will have more incentive to remain in the United States, and hence our country will retain more good-paying jobs.

Build a Tax System

Another area that our country could look into to see how we could create more racial/ethnic equality is how we tax people (again, recall our discussion in Chapter 3 on decreasing inequality). African Americans, Hispanic Americans, and Native Americans are disproportionately poor. During the last 30 years of the 20th century, these three groups typically had poverty rates of approximately 30% of their respective populations, whereas whites had poverty rates of approximately 10%. Given that racial/ethnic minorities are disproportionately poor, we could decrease various taxes on poor people, the near poor, and the working poor, and this would help them to keep more of their take-home pay and hence help them to get out of their poverty. This action would be a step forward for many African Americans, Native Americans, Hispanic Americans, and other racial/ethnic minorities who are disproportionately poor.

Another way we could change the tax system so that poor minorities and poor whites could become more equal to nonpoor people in our country would be to change the way we tax people on their social security. As we discussed in Chapter 4 on poverty, people who have jobs pay 6.2% of their wages in social security tax if they earn $94,200 or less per year. However, people who earn more than $94,200 do not pay social security tax on the money they make above this amount. We could create more racial/ethnic equality by having a progressive social security tax where poor and near-poor Americans pay a lower percentage in social security tax. Because a higher proportion of minorities have poverty and near-poverty incomes, they will have more take-home pay and hence not have need to live so close to the margin of subsistence. As a result, we, as a country, will have more racial/ethnic equality.

Another way to create more racial/ethnic equality is to provide a larger earned income tax credit, where the federal government gives additional money to people whose incomes are below the poverty line. We could continue to do this and increase this amount so that poor people who are racial/ethnic minorities, as well as poor whites, receive enough money to moves them above the poverty line. Gans (1995) predicted that our country might need to provide more income for people in this way if our economy does not provide enough income from jobs or does not provide enough jobs for everyone. Because African Americans, Hispanic Americans, and Native Americans are disproportionately poor, increasing the earned income tax credit would increase their incomes and help our country to achieve more racial/ethnic equality.

Provide More Social Services

If our country could provide more funding in areas such as (a) Section 8 subsidized housing for the poor or near poor (recall our discussion of Section 8 in Chapter 4 on poverty), (b) more public transportation to allow many of the poor who cannot afford personal transportation to get to jobs (also, recall our discussion on this matter in Chapter 4 on poverty), and (c) more subsidized health care (see Chapter 10 for an extended discussion), these services, along with others, would disproportionately help minority groups and poor whites to move us in the direction of more racial/ethnic equality.

In other words, any way we can help poor Americans to increase their standard of living, and hence create less inequality in our country overall, will also help minority Americans to have a higher standard of living and thus create less racial/ethnic inequality. We all know that increasing funding for these social services will mean that more tax revenue will need to come in to pay for these services. And this means that some people will need to be taxed more to pay for these programs. As I stated previously, the least sacrifice would occur if we increased taxes on those with the highest incomes and the most wealth. Those Americans with the highest incomes and the most wealth would still retain much of their incomes and wealth and hence would still enjoy a very high standard of living relative to the vast majority of other Americans.

I conclude this chapter with the following thoughts. Our American history has been filled with much prejudice and discrimination; there is still leftover prejudice and discrimination within some Americans even today; institutional discrimination, such as the policy of local property taxes funding public schools and the policy of "last hired, first fired," continues to work to the disadvantage of minorities; and the historical effects of prejudice and discrimination have resulted in minorities living disproportionately in geographic areas of high unemployment, leading to higher rates of poverty, homelessness, and stress in the family, and living in areas of higher crime rates. Given these social and historical factors, our country could implement a number of the preceding policies as a way to achieve more racial/ethnic equality.

Questions for Discussion

1. Should race/ethnicity continue to be used as one of the variables that undergraduate and graduate schools use in making admission decisions?

2. What variable or variables should be used to decide who gets admitted to a college or a graduate school? For example, should any of the following be considered as a legitimate variable: if one is an athlete, if one is the son or daughter of an alum, if one is a male or female, if one plays a band instrument or is a singer, if one is an artist, if one comes from a family whose members have given a lot of money to the school, or if one is a bright student?

3. Where could our country go from here in terms of achieving racial/ethnic equality?

4. Where do you think our country should go in terms of achieving racial/ethnic equality?

5. What do you predict will happen with racial/ethnic inequality during the next 10 to 20 years? What is your reasoning?

6. What do you think will be better for our country in terms of more unity in the country: Should we promote more interracial marriages, or should we promote more of racial/ethnic groups marrying within their respective groups?

7. How would you characterize the current racial/ethnic inequality in our country?

8. Which suggestions in this chapter will be tried, and which ones will not? Why?

9. What are your predictions for groups such as the Ku Klux Klan?

10. How do you think the issue of illegal immigration can or should be solved?

6

How Can We Solve the Problem of Gender Inequality?

There is less inequality between women and men than there was 50 years ago. However, there is still a gap measured in a number of ways. For example, during the early 1960s, "women earned about 59 cents for every dollar men earned" (Armas, 2004, p. A1). According to 2000 census data, women earned approximately 74 cents for every dollar men earned (p. A1). If we look at specific kinds of jobs, say medical doctors, men earned $140,000 per year, whereas women earned $88,000 per year (63% of men's income). Male lawyers earned $90,000 per year, whereas female lawyers earned $66,000 per year (73% of men's income). As for chief executives, men earned $95,000 per year, whereas women earned $60,000 per year (63% of men's income) (p. A1). As you can see, we are moving toward more gender equality in terms of yearly income, but we have not yet reached equality.

Causes of Continuing Income Inequality

Some of this income gap is due to women getting into more professional fields during recent years and not yet reaching the senior level in their careers. Also, some of this difference is due to some leftover prejudice and discrimination against women. A third reason for the difference is that a number of women will choose certain kinds of occupations, such as certain specialties in medical practice, law, and chief executive positions, that pay less than other kinds of positions in those respective fields. However, as women disperse throughout the various kinds of positions within various occupations and move their way up to senior-level positions, the gap between their incomes and men's incomes will continue to decrease, thereby moving our society toward more income equality between women and men.

What Can We Do?

If we want our society to move toward the goal of gender equality where women and men are treated equally, have equal opportunities to get as much education as they want and to get the wide variety of jobs for which they are prepared, have equal chances as do men for promotion and upward mobility, and have the chance to gain the same amounts of money, power, and prestige as do men, there are a number of actions we can take.

Socializing Girls and Boys More Similarly

One of the most important actions we can take is to socialize children at home and in school to treat both boys and girls equally. For example, in terms of socialization in the home (with *socialization* being defined as people learning the ways of a society, e.g., norms, values, beliefs), parents can socialize or teach their daughters to play sports, play with mechanical things, play with toys such as bulldozers and trucks along with playing with dolls, read about girls and boys doing a variety of jobs, and watch television shows that portray both girls and boys as well as men and women being active, making decisions, and doing a variety of activities. To the degree that parents socialize their children in this manner, children will think that it is "natural" for both girls and boys to do a wide variety of activities.

Recall in our theory of conflict and social change (see Chapter 1) that we, as humans, socially construct our norms, values, and beliefs. Some of these norms, values, and beliefs have to do with what males and females are supposed to do. For thousands of years in many cultures, people socially constructed the idea that males did certain things and females did certain things. There was typically some overlap of what males and females did, but there was always some variation, and sometimes a lot of variation, in what males and females did depending on the culture. Since the late 1960s with the rise of the modern women's movement, we have begun, in our society, to increase the overlap of what males and females do. As we have gone in this new direction, we have been in the process of changing how we socialize males and females. More and more, we are teaching or socializing males and females to do the same things such as play sports, get more education, and work in the same kinds of jobs.

Because we are already going in this direction in society, this leads to the following question: How can we socialize more parents to socialize their children in the previously mentioned way as a means to create more gender equality? One additional step that our society could take is to create parenting classes in elementary, middle, and high school that teach students not only how to cook, clean, pay the bills, and feed children nutritious food (for more on this, see Chapter 10 on health care) but also how to give their future children equal opportunities. In other words, a key part of the parenting class would be to teach parents how they can raise girls and boys to have a variety of careers rather than to teach girls to raise the children, wash the clothes, cook the food, and clean the house and teach boys to seek more education, make the family decisions, and want jobs and professional careers.

Schools can do other things to promote more gender equality. For example, teachers can treat boys and girls equally in the classroom. Many teachers are already becoming more aware of how they can treat girls and boys equally. The proportion of times they call on boys or girls, whom they praise and reinforce, and how they praise and reinforce—all of these teacher-controlled actions are ways they can consciously work to make sure they treat both genders equally. Teachers typically have some freedom in picking the reading materials for their students. They can intentionally select reading materials that show both men and women being active, making decisions, and working at a variety of jobs where these jobs show both men and women occupying leadership positions with considerable money, power, and prestige. In other words, teachers have the ability to be a major influence in promoting more gender equality.

To teach more gender equality, teachers will need to be trained accordingly. College methods classes can show future teachers what kinds of methods and techniques they can employ and what kinds of behaviors they can display to promote gender equality. Likewise, college and university professors in courses other than education courses are already becoming more aware of treating both genders equally and are taking steps to promote more gender equality such as becoming more conscious of calling on female and male students equally and grading exams blind so that there can be no gender, racial, or other kinds of discrimination. Also, colleges and universities in general are hiring more female professors, promoting them to associate and full professorships, appointing them to be department heads and deans of faculties, and choosing them to be college and university presidents.

A very important part of creating an atmosphere of gender equality in elementary, middle, and secondary public school classrooms is to elect members of school boards who are sympathetic to the idea that females and males should be treated equally. Parents' organizations can be influential in helping to elect these kinds of people to the school board and can make their voices heard. The more people in public education (e.g., school board members, superintendents, principals, teachers, coaches, counselors) desire to create equal opportunities for female and male students in the public schools, the more likely they will achieve this goal.

In many small communities, one individual (or one group or organization) with strong beliefs about this issue can make this issue visible by making his or her (or its) voice heard at school board meetings and parent–teacher association meetings. In other words, individuals, groups, and organizations can act as external pressures to promote more gender equality in schools. This idea of promoting more gender equality has never been viewed as favorably as it is now. Many Americans are realizing that if we want to fully embrace our cultural value of equal opportunity, we must apply it to both genders.

Otherwise, the custom of "this is the way we have always done things" will continue. The result will be that we will continue to have a social structure that reinforces gender inequality instead of gender equality. The old adage that "the squeaky wheel gets the grease" seems to apply to this situation. If so, this is an instance where individuals, groups, and organizations can make a difference with respect to a particular social problem in our country.

Subsidized Child Care

To move toward more gender equality in our country, one important thing we could do is to provide more child care that is affordable, safe, accessible, and developmental (Crone, 1998). The reason why this is important is that traditionally women have borne the major share of raising and taking care of children in the family. So, as women get out into the occupational world and seek to have careers where they have a chance to be as upwardly mobile as men, they will need to have child care available to them. Otherwise, women with children will continue to be at a disadvantage, compared with men, in developing their careers and therefore being upwardly mobile.

For poor and near-poor mothers who work, not only finding safe, accessible, and developmental child care is important but also finding child care that is affordable is a big challenge because paying for child care can take a lot of a mother's wages. One of the key ways for mothers to have more money is to receive some kind of subsidized child care, especially for poor and near-poor working mothers. Currently, our federal and state governments provide for some subsidized child care, but there are many working mothers who have applied for subsidized child care but do not receive any help.[1] This means that many of these working mothers might need to quit their jobs because they cannot afford to pay for child care—sometimes as much as $230 per week (Yetter, 2003, p. A4). In other words, if we want poor and near-poor mothers with children to have jobs and survive financially, to get off and stay off of welfare, and to have a chance of being upwardly mobile, we will need to help them with more subsidized child care. As women can get into the workforce and stay in the workforce, women will make more money and have chances to be more upwardly mobile, thereby creating more gender equality.

This raises the question of how we use our federal tax money and how much tax money we raise. This becomes a political question, but I want you to become aware of this question. Congress and the president decide how much to tax, whom to tax, and how to use the tax revenues they receive. Should we bring in more or less tax revenue? Should we tax the poor, working, and middle classes more, or should we tax the rich more? Should we spend the tax revenue the government receives more on the military, health care, child care, or some other area? These are the decisions that Congress and the president make—huge, crucial, and extremely important decisions for all of us. So, the question of paying for more subsidized child care must be placed within the larger context of how we will spend our tax money and how much tax money we decide to raise.

As you recall when we talked about conservatives and liberals in Chapter 2, conservatives tend to want to have less spending on social services and more spending on the military, as the conservative presidencies of Reagan and Bush (junior) have demonstrated. Hence, a main barrier to allocating more spending of tax dollars for subsidized child care will probably meet with more resistance from conservatives and more acceptance from liberals. If conservatives are in the White House and control both houses of Congress (as has been the situation during the first years of the 21st century), and if there is considerable deficit spending (i.e., the government is spending more than it is collecting in taxes) due to tax cuts and a war occurring at the same time, there is a smaller likelihood that there will be increased spending for subsidized child care. As you can see, the prospect of increased spending for more subsidized child care as one method of providing more opportunities for poor and near-poor working mothers to get and hold jobs and have the potential for upward mobility and hence create more gender equality does not look promising at this time in our country.

More Women in Local, State, and National Public Offices

In all probability, as more women run for political offices and win local, state, and national offices, both liberal and conservative, their presence in legislative bodies will be a major factor in moving our country toward more gender equality. I include conservative women, especially politically moderate and more educated conservative women, because they will be more insistent that women be paid the same as men for doing the same jobs, that women be given the same opportunities in various careers as are men, and that women be given the same chance for upward mobility as are men.

If, on the other hand, more extremely conservative women and men, such as fundamentalist Christian conservatives, get elected to political offices, the idea of gender equality in general, and of child care subsidies in particular, will probably not expand much and could even diminish. During the coming years, while wearing your "sociological hat," you might analyze the relationship among fundamentalist conservatives, moderate conservatives, and liberals and how they address the social problem of gender inequality.

Good-Quality Public Schools

A key factor in achieving more gender equality in our country is to have good-quality public schools. Such schools will allow all Americans to be better prepared to go to college or trade schools so that they can get better paying jobs. Currently, with a number of children going to public schools that do not adequately prepare them to go to college or trade schools, these people are more likely to end up taking minimum wage types of jobs that leave them below the poverty level in our country with little chance of being upwardly mobile. Women in general, and female heads of households in particular, who are poor and have little education and few skills are in a situation where they are less likely to be upwardly mobile. To work toward more gender equality, we will need to get all children better educated so that they can have a better chance for more upward mobility. Consequently, excellent public education (for a more detailed discussion of how we can create an excellent public education system for all Americans, see Chapter 7 on education) and training not only will allow Americans in general to be more upwardly mobile but also will allow more females to develop the educational tools to get ahead, thereby promoting more gender equality.

Women Becoming More Indispensable to Society

To move toward more gender equality, it seems imperative, according to Rae Lesser Blumberg, that we take steps to make sure women become more indispensable to our society.[2] As women become more indispensable, they will be in greater demand in the workplace. With this greater demand, they will have more economic and political power in society. With more economic and political power, they will be able to have more influence over the creation of laws and social policies that will work to their vested interests. For example, women will have more influence over laws regarding their rights in marriage and divorce, their rights in deciding whether or not they want to have children, and their rights in deciding when they want to have children via laws that allow for the use of the birth control pill and the "morning after" pill, the right to have an abortion, and other kinds of technologies that give women more decision-making power over their lives.

In the past, under the more traditional gender role by which both men and women were expected to abide, women had little or no power over the conditions under which they could have sexual relations, the conditions under which they could have children, and the number of children they could have. As women have gained in economic and political power, however, they have been able to influence the passing of laws that implement social policies that work to their vested interests and hence move our society toward more gender equality.

A key to this process will be that women's labor needs to be made more indispensable, which will give women more economic and political power, which in turn will help them to pass laws and create social policies that will create more gender equality. For this general process to happen, initial steps such as good-quality education for females and more child care subsidies for working women will be needed as an essential base for women to become more indispensable.

Taxing and Gender Equality

If we tax poor and lower income people less, they will have more money left over from their paychecks to pay their bills, allow their children to go to school to get more education and training, and hence enable their children to be more upwardly mobile in our society. If we, for instance, lowered the income tax rate on lower income people and lowered the social security tax on them as well (and at the same time raised the income tax rate and social security tax on the wealthiest people, say the wealthiest 1%, 5%, or 10% of the people in our country), lower income people in general, and lower income female heads of households in particular, would have more income. There would be less income inequality in our country in general, and there would be more income for poorer women in particular, moving lower income women in the direction of more gender equality.

Health Care and Gender Equality

Providing health care, at first thought, might not appear to be related to achieving more gender equality. On further reflection, however, it may also be of great help to women gaining more gender equality (for more details, see Chapter 10 on health care). Currently, the United States is the only industrialized country that does not have a health care system for all of its citizens (Kerbo, 2003, p. 36). If we had a health care system for all Americans, this would greatly benefit lower income female heads of households. Lower income female heads of households would, in a sense, have higher incomes because they could spend the money they earned from their jobs for other things besides health care. Health care, along with lowering income and social security taxes on lower income people, would help female heads of households, especially poor and working-class female heads of households, to move in the direction of more gender equality.

Paid Maternity Leaves and Gender Equality

A key factor related to providing health care that would move our country toward more gender equality is providing paid maternity leaves for women. Currently, American women can take 3 months off from work, but their maternity leaves are unpaid, unlike the situation in many European countries that provide paid maternity leaves.[3] As a result, many American women of moderate and lower incomes cannot afford to take maternity leaves that are unpaid. Providing paid maternity leaves for women would especially help poor, near-poor, and working-class women who could then take maternity leaves and hence provide better for their babies. These paid leaves would also show that our society is considerate of the needs of mothers in the workplace. These mothers would be able to maintain their incomes and work statuses in their jobs, thereby allowing them to be as upwardly mobile as men. Providing paid maternity leaves for women in the United States would move our country in the direction of more gender equality.

Social Services and Gender Equality

As a side note, you should be aware that usually when there are more social services for Americans in general, such as social security for the elderly, health care for all Americans, child care subsidies for poor and working mothers, good-quality public education for all children, and paid maternity leaves for all women, these social services, taken together, will mean that there will also be more services for women in particular. As women have access to more social services, our country will move in the direction of more gender equality.

Minimum Wage and Gender Equality

Another way for women to move toward more gender equality is for the United States to increase the minimum wage. Many mothers in the United States work full-time for $5.15 per hour and are still thousands of dollars below the poverty line. Christopher (2004) noted,

> When working full-time (at least 35 hours per week), about one-third of U.S. women and more than 40 percent of U.S. single mothers earn wages too low to free their families from poverty. In other Western nations, particularly Sweden, The Netherlands, and the United Kingdom, working full-time pulls the vast majority of women (including single mothers) and their families above the poverty line. (p. 108)

As you can see, raising the minimum wage in our country would be especially helpful for women in our country so that they could make, at the very least, poverty-level wages instead of making wages considerably below the poverty level. Yes, this may result in some job losses, but based on the 1990–1991 and the 1996–1997 minimum wage increases, the job losses would not be "significant" (Christopher, 2004, p. 111).[4] If we want to help lower income women to increase their incomes and thus help our country to move toward gender equality, an increase in the minimum wage could be of help to many women.

Abuse and Gender Equality

Historically, the use of violence against women in the home has been a barrier that has kept women "in their place" and hence has perpetuated gender inequality. As our society can decrease violence against women in the home, these women will be more comfortable in speaking up for what they think is right and speaking up for their vested interests. A consequence of their being able to speak up more will be to move our society toward gender equality.

Ways to decrease violence in the home are varied. Socialization in public schools via parenting classes, where boys are socialized to be husbands and fathers who use discussion and negotiation, rather than physical force, as methods of interacting with wives and children, is one way (for more details, see Chapter 11 on families). Another way to decrease violence against women is to make safe houses and shelters available for women and children to go to whenever violence occurs in the home. Also, the option for women to take their violent husbands to court to get restraining orders put on them and to get their violent husbands to serve jail time are other methods that can serve to deter violence against women. Also, as women take more positions within the criminal justice system (e.g., police officers, probation officers, prosecuting attorneys, defense attorneys, judges), women will have more understanding and sympathetic support from the criminal justice system than they did in the past when men made up the entire criminal justice process and were not as understanding or sympathetic. This social change will promote more gender equality in our society.

How Men Are Taught to View Women and Gender Equality

Likewise, our country can change the way we socialize boys and men, where instead of women being portrayed as sex objects in ads, movies, and television shows viewed by boys and men, women can be seen in other ways. Currently, we are somewhat conscious of working toward the goal of seeing and treating women in other ways. However, a quick look at current ads on television, in magazines, and along highways suggests that we have a long way to go. As we socialize boys and men to treat girls and women in what we might call *multidimensional* ways (i.e., boys and men are taught to see girls and women in many different ways rather than seeing them only as sex objects), we should see a subsequent decline in rape and sexual harassment against women. As teachers treat both boys and girls as multidimensional individuals, as schoolbooks portray boys and girls in these multidimensional ways, and as parenting classes teach future husbands and wives to treat each other as multidimensional persons, we will begin to create a new generation of males and females who see each other in many ways and not just in a one-dimensional, sexual object type of way. During the last 30 years of the 20th century, we began to be conscious of males being taught by our culture to see women mainly as sex objects and began to change this perception, but we will need to continue to work on this during the 21st century as we move toward more gender equality.

We will face a number of dilemmas as we move toward a society where women are seen in a multitude of ways. Do we ban pornography—in other words, censor what adults can watch? Do we ban ads that constantly use sexual attraction of males toward females as a way to sell products? Do we ban women from dressing scantily in terms of showing parts of their breasts, buttocks, and midsections? Do we socialize women not to flirt with men? Do we socialize boys and men to get them not to think sexual thoughts so often? To what degree can and should we socialize sexual thoughts out of males, and to what degree should we confront biological limits of suppressing sexual desire in men? There is a need for further inquiry, research, and discussion on a number of these matters.

Currently, our culture is sending mixed messages. On the one hand, we are trying to move toward seeing women in multidimensional ways. On the other hand, we are still putting a lot of emphasis on women being seen as sex objects in ads, television shows, and movies. So, as a society, we still have a way to go before we have worked out this aspect of gender inequality.

Equal Rights Amendment and Gender Equality

Another way we could move toward gender equality in our country is to pass the Equal Rights Amendment, which states that people cannot be discriminated against based on their gender. Congress passed this amendment during the early 1970s, but not enough state legislatures ratified it. A number of people might question whether an amendment to the U.S. Constitution is needed when we already have a number of laws and U.S. Supreme Court decisions favoring more equality for women, including passage of the Nineteenth Amendment to the U.S. Constitution in 1920, where women received the right to vote (Eitzen & Zinn, 2003, p. 262); passage of the Equal Pay Act of 1963, which prohibited employers from paying women less than what men are paid for doing the same job; passage of the Civil Rights Act of 1964, which "banned racial, ethnic, and sexual discrimination in employment and union membership" (Farley, 1995, p. 306) and "prohibited discrimination by privately owned businesses providing public accommodations" (p. 306); passage of Title IX Act of 1972, which required "schools that receive federal funds to provide equal opportunities for males and females" (Eitzen & Sage, 2003, p. 319); and the 1973 decision of the U.S. Supreme Court in *Roe v. Wade,* which allowed women to decide whether or not to have an abortion. So, why have a constitutional amendment when the rights of women already seem to be covered?

There are two reasons. One is that there still might be areas of discrimination that the preceding laws do not cover. Such an amendment to the Constitution would give women the legal right to go to the courts to seek a redress of grievances if there was some kind of unequal treatment based on a person's gender. A second reason is the symbolic value that such an amendment would give. That is, such an amendment would state, in a very clear and visible way, that our country intends to treat women and men equally.

Concluding Thoughts

A concluding note on what we can say as the preceding policies are carried out in our country is that as more women are given equal opportunity, this will mean that our society will receive more of the fruits of the contributions of women who make up one half of our population. As a result, our society will become more efficient in that more Americans will be able to live up to their potential. Adding each individual contribution to the whole of the society will make for much more contribution to society.

Yes, the level of competition will rise as more women get a chance to "throw their hat into the ring." And yes, a number of men will lose out to women who are better qualified and have higher skill levels. These consequences will occur whenever a society gives equal opportunity to a new group of people, as we have seen, for example, in the case of African Americans in sports. Such is the nature of a society where everyone is given equal opportunity.

There is a good chance that many, if not all, of these suggestions for creating more gender equality will eventually be implemented. The reason why I say this is that since the 1950s, with the onset of the civil rights movement by African Americans followed by social movements of women, Native Americans, gay and lesbian Americans, handicapped Americans, and elderly Americans, there has increasingly become in our society a spirit and a consciousness of moving toward more equality along a number of dimensions of life. It seems that we have therefore created a momentum that will continue to move our society in the general direction of overall equality and in the specific direction of gender equality.

Questions for Discussion

1. In addition to what we discussed in this chapter, what else can we do to move toward more gender equality in our country?

2. What do you predict for the next 5, 10, and 20 years with respect to gender inequality and equality in our country? What is your reasoning?

3. What will be key barriers toward our moving toward more gender equality?

4. What are steps we can take to overcome these barriers?

5. What are the various positives or advantages that favor our country moving toward more gender equality?

6. Will we have a woman as president? What is your reasoning?

7. Should women serve on the front lines in battle in the military? What is your reasoning?

8. If we had an amendment to the U.S. Constitution with regard to gender equality, what might be some consequences of this amendment?

9. Given that this chapter focused on the United States, what do you predict will be done at the world level with respect to gender equality?

10. What do you think should be done at the world level with respect to gender equality?

7

How Can We Solve the Problem of Unequal Education?

A number of elementary, middle, and high school students in our country go to very good schools. They have well-qualified teachers, decent-sized classes, good facilities, and a safe environment. They are getting the kind of educational experience that we hope all children can get. However, as you no doubt know, many children in our country do not go to these kinds of schools. Kozol (1991), in his book *Savage Inequalities*, documented some of the deplorable conditions that a number of our American children must endure in going to school to get an education. Moreover, with the economy being in bad shape during recent years, and with the local, state, and national governments having difficulty in paying their debts, money for schools is tighter. Class sizes go up, and services such as counselors and extra teacher assistants are cut (Rodriguez, 2003). So, here we discuss some things we could do to make excellent schools for all American children.

What Can We Do?

Certified Teachers

We could hire only teachers who are certified to teach in the areas they are hired to teach. Currently, school systems will hire people who are not certified to teach what they are assigned to teach. As a result, students who have these teachers are at a disadvantage compared with those who are taught by certified teachers.[1] Also, at times, even certified teachers are required to teach in areas where they are not certified. This hurts their morale and enthusiasm.[2] They spent years preparing to teach one subject, but they do not get to teach it. So, not getting certified teachers or not placing teachers in the areas of their certification hurts both students and teachers.[3]

To get more qualified teachers who are certified in specific areas to teach, and to have enough of these teachers for all public schools from Grade 1 through Grade 12, we will need to carry out two measures. First, we will need to have more stringent requirements as to who can teach. The problem with this is that schools cannot always find certified teachers to fill certain positions, so they temporarily "fill in" with people who are not certified to teach the subjects they were hired to teach. They may be conscientious, smart, and good people, but they do not know the subject matter in much depth (if in any depth at all). For example, a school might need a certified teacher in biology. However, the school cannot find such a teacher, so it does the best it can and hires, for example, a college graduate who has never taken any biology courses in college or maybe has taken one or two such courses. Assuming that this person does the best he or she can, this person not only does not have the depth of understanding of the subject matter but also does not have the love of the subject matter. Furthermore, this person has no grounding in teaching methods. The students who have these kinds of teachers are at a disadvantage in their education and hence in their future chances for upward mobility.

One way to get more certified teachers is to pay teachers more money. Higher pay will attract more people into teaching and provide more certified teachers for our public schools.[4] Many people who could be excellent teachers will not consider going into the field of teaching because it pays lower salaries than what other college-educated occupations pay. Once we get more certified teachers, we can fill our classrooms with well-qualified teachers who love to teach.

Class Size Reduction

A second measure we could take is to reduce class size.[5] By making classes smaller, the teacher can give more individual attention to each student, and this will result in better test scores.[6] Research suggests that smaller classes help students to complete high school, graduate on time, and graduate with honors.[7] The National Education Association (NEA), a group that promotes better public education for American children, supports a class size of 15 students in regular programs and an even smaller class size in programs for students with exceptional needs. Teachers with small classes can spend time and energy helping each child to succeed. Smaller classes also enhance safety, discipline, and order in the classroom. When qualified teachers teach smaller classes in modern schools, kids learn more. It is common sense, and the research proves that it works to increase student achievement.[8]

A low teacher-to-student ratio is especially important to kids from low-income families who have not had a chance to get much help from their families. For example, they have not had the chance to travel; go to museums; have books, magazines, and newspapers around the house; own and use a computer; or get knowledge about college from their parents because their parents are less likely to have had a college education.[9] So, low-income children, more than middle- and upper middle-class children, will need more personal attention in the classroom to make up for what they might not get from their family environments.

Up-to-Date Facilities and Equipment

To have excellent public schools, we also need to have good, up-to-date facilities and equipment.[10] Children who study using out-of-date facilities and equipment are at a disadvantage.[11] Different parts of the school, such as restrooms and labs, may be rundown and not work. Paint may be falling off the walls, and the facility might not be a place where students (and teachers) are proud to go to school. Children might not have running tracks, football fields, gyms, choir rooms, or band rooms where they can practice or play. They might not have microscopes, band instruments, band uniforms, computers, textbooks, chalk, and paper, all of which are needed in the contemporary school to give students a full education.[12] Jonathan Kozol observed the following about the schools in East St. Louis: In the boys' bathroom, four of the six toilets did not work, there was no toilet paper, and there were no toilet seats (Kozol, 1991, p. 34); a history teacher had 110 students but only 26 textbooks (p. 35); there were windows without glass, and there were dark hallways because there were no light bulbs in the sockets or light bulbs had burned out (p. 36); certain classrooms were so cold in the winter that students needed to wear their coats (p. 37); the physics lab had no equipment (p. 30); the high school had no videocassette recorders (VCRs) (p. 29); the biology lab had no lab tables (p. 28); there were few dissecting kits, and even the ones available were incomplete (p. 28); chemical supplies, even in a city that had two chemical plants, were scarce (p. 28); "I need more microscopes," said a teacher (p. 28); and so on.

With such deplorable conditions in terms of facilities and equipment, it comes as no surprise that the children in the East St. Louis schools were not well prepared to get jobs or go to a vocational school or college. These public schools did not prepare the children to become independent adults. It therefore seems likely that these children (and other children in similar social conditions throughout the United States) will have higher rates of unemployment, poverty, homelessness, and crime.

This situation is a good example of the application of the concept of the *sociological imagination*, where we realize how the larger social structure in which people live can cause them to have personal troubles (Mills, 1959). Having inadequate public schools will lead to more personal troubles, which in turn will lead to more negative consequences. If we, as a society, do not want these negative consequences (see Chapters 8 and 9 on crime and drugs, respectively), we will need to create better schools.

Smaller and Safer Schools

To have excellent public schools means that the schools are safe to attend.[13] Part of having safer schools is having more teachers, as discussed previously. Also, part of having safer schools is having smaller schools[14] so that each student is seen as "somebody" who is more likely to feel integrated within the school.[15] With fewer students, a higher percentage of people at the school can be on teams, in choirs, in bands, in clubs, and so on. A smaller school means that a higher percentage of students are able to find their "niche" within the daily offerings of activities that the school has to offer. As a higher percentage of students becomes more involved and committed to the school via its various activities, there will be less likelihood of these students gravitating toward illegitimate activities. So, having smaller schools with various activities available, where children feel as though they are a part of the school, should act as a major deterrent to illegitimate activities—both in and out of school.

Problems That Prevent Us
From Having Excellent Public Schools

To have smaller schools, this means that we will need to build more schools. And, of course, this means that we will need to spend more money on education. Most Americans would like to have all American children have a chance to get a good public education. Yet, at the same time, many of these Americans feel as though they are taxed enough and therefore do not want to create a better public education system if it means paying more taxes. Moreover, because much public education (Grades 1–12) is paid for by local property tax, there is little desire by many people to pay more property tax to pay for better public schools given that they already pay many other taxes (e.g., federal income tax, state income tax, state sales tax).

Another problem is that some people choose to pay extra money to send their children to private schools. They do not want to pay any more property tax that goes to pay for public schools when they send their kids to private schools. So long as their children are taken care of educationally, why should they need to pay for the education of other people's children?

A third barrier to creating excellent public schools is that the federal government and state governments are currently having financial problems. After running surpluses during the late 1990s, the federal government is spending more than it receives in tax money (called *deficit spending*). Moreover, 44 of the 50 states are having major financial problems (NEA, n.d. b).

So, what can we do to bring about the possible solutions discussed at the beginning of this chapter that will provide a good-quality education for all Americans in light of these barriers?

Tax Money for Schools

To have good-quality schools, we will need to invest more money in our public schools to get more certified teachers, have smaller classes, and have smaller schools that are well maintained and have the needed facilities and equipment. It follows that we need to have more of our tax revenues funneled toward our public schools. Fruchter (1998) asserted that to achieve this, we need to reform "the way we fund public education" (pp. 15–16).

Instead of placing this added tax burden on property owners, a more humane and realistic way could be to create a more progressive federal income tax where the wealthiest Americans, say the wealthiest 1% to 10%, pay more federal income tax. Because the wealthiest 1% of the people in our country own 38.5% of all the wealth, and the wealthiest 10% of Americans control 71.8% of all the wealth (Kerbo, 2003, p. 31), they are the most able to pay more tax and still live a good lifestyle.

The tax money that is collected by the federal government from the wealthiest 10% of people in our country could be distributed to local school districts with the stipulation that the administrators of these local districts use this money in four areas: (a) hire only certified teachers to teach in their certified areas; (b) hire more certified teachers to reduce the class size to 15 or fewer students in low-income school districts; (c) build and maintain facilities such as band rooms, choir rooms, libraries, auditoriums, gyms, tracks, and fields and also buy the needed equipment such as microscopes, band instruments, choir music, pianos and organs, computers, and books; and (d) build more schools to create smaller schools so that a larger percentage of the student body can be in various activities and thus feel like they are a part of their school. By spending tax money in these kinds of ways, we can more fully ensure that school districts will create good-quality schools.

Local school districts and local school superintendents, school boards, principals, teachers, and citizens of that district will still decide whom to hire as certified teachers, where to put kids in what classes, what classes are required, what overall curricula are created, what books to order, how to build the facilities and what the facilities will look like, when the school year starts and ends, what holidays will be taken, and what specific equipment to buy. In other words, local people will still have much freedom to decide how they want to run their schools so long as they hire certified teachers, hire more certified teachers, have smaller class sizes, have smaller schools, and have well-maintained schools with up-to-date equipment.

Positive Consequences

If and when we create a good-quality school system in our country, there will be numerous positive consequences that follow. When we can provide an excellent education for all American children, these children will be more prepared to be upwardly mobile in our society. As young people are more prepared to be upwardly mobile, they will experience less frustration in a society that tells them to be upwardly mobile and yet does not give them the means to be so.[16] Consequently, with a good education, young people will believe that they have a chance to move up in the society using legitimate means to be upwardly mobile.

When people believe that they can be upwardly mobile via legitimate means, they are less likely to believe that they need to resort to illegitimate means. Merton (1968) discussed this process in his article titled "Social Structure and Anomie." In this article, Merton noted that if a social structure that blocks opportunities and does not allow people to be upwardly mobile through legitimate means is created, people are more likely to use illegitimate means such as robbery, burglary, selling illegal drugs, and prostitution.

As we can give all young Americans a good education, we will give them the capacity to be upwardly mobile through legitimate means. As a result, we should see fewer people resorting to illegal means to get ahead and thus should see a decrease in certain kinds of crime (the focus of Chapter 8). We should also see decreases in the costs related to crime such as the need for fewer police, less prison space, less prison personnel to look after prisoners, and less cost in court time. We can then allocate this tax money for other purposes.

Moreover, if all young poor Americans receive an excellent education, they have more of a chance to get out of their poverty. The result will be that we, as a country, will not need to spend as much tax money on poverty programs to help people survive because they will be able to survive on their own. Also, poor and near-poor children will have more hope that they too can get ahead in this world. They will also go to safer schools, and this will be a more fulfilling educational experience for them as well as a great relief to their parents.

Another positive consequence of providing good schools for all American children will be that the reality of giving people an opportunity will be closer to our ideology that says that everyone should have an equal opportunity. Some people think that we already have equal opportunity in our country for all young Americans. However, we know that a person's social class, race, and gender still play a role in who has more or less opportunity in our country. With an effective education for all American children regardless of their class, race, or gender, our country will be closer to living up to its ideology of providing equal opportunity for all Americans.

What Will Not Work if We Want Excellent Public Education: The Voucher System

President George W. Bush's voucher plan will not create good-quality public education for all American children. With the voucher plan, a child can go to another public or private school, and that school will receive a voucher—a certain amount of money (e.g., $3,000)—to pay for that child coming to the school.[17] The intent is to give children, especially poor children going to deplorable schools, the chance to go to good schools.

There are specific problems with this policy in addition to more general problems. As for the specific problems, just because a poor child may want to go to another school does not mean that the other school wants that child as a student or has room for that child even if the school officials are willing to accept that child and other children like him or her.[18] Ask yourself this question: If you were a school official and were faced with 10, 20, 50, 100, or 200 children who wanted to come to your school as new students; these were poor and minority children who were behind other students in your school with regard to reading, writing, and math abilities and therefore would need extra attention to help them catch up; and your school was a predominantly middle-class or upper middle-class school with nearly all white kids, would you want to take on more students who are likely to be ill-prepared and from a lower social class?

Probably, a number of you would say "yes" in that you believe it is the right thing to do to give kids a better chance at a good education. But the reality is that this situation is a disincentive for many school officials to accept more children if a school is already filled or nearly filled and if the children applying to come to the school will require disproportionately more attention from the existing teaching staff.[19] This means that the existing teaching staff, facilities, and equipment will be stretched further. Also, what about the potential problem of racism and problems between white kids and black kids in the school? And what about the potential problem of different social classes not understanding or accepting each other?

These are questions that will be considered by officials at the school to which low-income kids are applying. I predict that a number of school officials, once they consider the consequences of accepting these new students, will turn down accepting a number of them. As a result, this policy will help a few low-income kids but will leave out most of them.[20] I predict that most low-income students will continue to go to the deplorable schools they are currently attending and that our public education system will not change much. Most of these low-income kids will still go to inadequate schools, get a poor-quality education, and not have the chance to be upwardly mobile in their lives and will instead end up working at minimum wage types of jobs or using illegitimate means to make money.

With respect to the voucher system, there is also the problem of taking the taxpayers' money and subsidizing children to go to private religious schools.[21] A number of people in our country are upset about this.[22] Because 85% of the private schools in our country are religious, there is a high probability that when a child uses a voucher to attend a private school, it will also be of a certain religious persuasion. Hence, we will use public funds to support private and religious education—a potential problem in keeping the church and the state separate (NEA, n.d. e).

In a more general way, the voucher system will not create a good public education system in our country for the following reasons. First, it does not do anything about getting certified teachers to teach in their certified areas. Second, it does not do anything about getting more certified teachers into low-income schools that really need these kinds of teachers. Third, it does not do anything about directing more tax money into inadequate schools. Fourth, it does not do anything about providing low-income schools with excellent facilities and equipment to allow certified teachers to teach well. Fifth, it does not do anything about creating smaller schools throughout the country so that more kids have a chance to take part in choirs, bands, clubs, and teams and therefore will feel more a part of their schools, thereby decreasing the chances of their resorting to illegitimate activities. None of these five conditions is changed by using the voucher system. We still have many poor kids taught by uncertified teachers in schools that have poor facilities and inadequate equipment, and we still have schools that are large, impersonal, and unsafe.

Moreover, this voucher system might end up being a subsidy for middle- and upper middle-class parents to send their children to private schools.[23] For example, instead of parents' needing to pay $7,000 to send their child to a private school, the voucher could knock $3,000 off the cost and allow the parents to pay only $4,000. The intent of the voucher plan is not to subsidize middle- and upper middle-income parents to send their children to private schools, but that is probably what will happen to a degree. By the voucher system enabling wealthier children to go to private schools at less expense to their parents, the result could be the unintended consequence of segregating lower social class children from higher social class children even more.

As for how the American public feels about the voucher system, "Most Americans oppose voucher programs" (Carroll, 2003, p. A9). Based on a national poll by Phi Delta Kappa International and Gallup, Carroll (2003) reported, "Support for a program that allows students and parents to choose a private school to attend at public expense dropped to 38 percent from 46 percent" (p. A9).

All in all, the voucher system will help some kids get a better education, but it will not help many children who really need the help and will therefore not create a good public education system for all children in our country. As Gutmann (2000) noted,

> There is no evidence that vouchers will produce good schools for the vast majority of children who need them most. This democracy—if it is to be worthy of the name—needs public action to create the schools that all our children deserve. (p. 24)

Questions for Discussion

1. How important is it that we have certified teachers?

2. How important is it to increase salaries for teachers to get more quality people to go into teaching? How could or should we pay for higher salaries?

3. What do you predict will happen with unequal public education in our country during the next 5, 10, and 20 years? Will it become more equal, stay about as it is now, or become more unequal? What is your reasoning?

4. If we had a national goal of making all public education of good quality, where could and should we get the money to pay for this?

5. How can the federal government help?

6. Should the federal government help?

7. What do you think should be the responsibility of local school districts and school boards in comparison with that of state governments and the federal government regarding education?

8. What would you do to make public education excellent for all American children?

9. How do you think we should pay for improving our schools?

10. How would you improve our public schools?

8

How Can We Solve the Problem of Crime?

A t first thought, you may think that I am going to suggest building more prisons and adding more police, investigators, judges, courtrooms, and probation and parole officers. No, I do not go in that direction to decrease the crime we have in our country. I go in another direction to decrease our crime, and I say more about this shortly.

You also may think that I am going to address all crime. No, I do not do that. I focus on the crimes that have largely, if not entirely, social causes and that are related to the influences of poverty, much inequality, and racial/ethnic and gender discrimination (as a lead-in to helping us understand the kinds of crime I address in this chapter, see Chapters 3–6). That is, how might the existence and perpetuation of poverty, much inequality, and racial/ethnic and gender discrimination lead to certain kinds of crimes, and what can we do about these crimes?

I want to focus on crimes that are more likely to occur in our society due to no opportunities, little opportunities, or blocked opportunities in our social structure. By this, I mean that there is something about the social structure of our society, either now or in the past, that has stopped or greatly hindered people from using the legal means of making a living, surviving, and getting ahead in American society. For example, from the early 1600s to the mid-1800s, our country had slavery. This was a social structure, that is, a social pattern of laws, informal norms, beliefs, values, statuses, and roles that were interrelated and persisted over time. As I emphasized in the theory of

119

conflict and social change in Chapter 1, humans create social structures. And yet, as Marx pointed out, these social structures can control humans in a number of ways, including that humans are not able to make a living, survive, or get ahead (Marx, 1964, 1967; Marx & Engels, 1992).

In the social structure of slavery, for example, African Americans, no matter how smart, creative, and hardworking they might have been in 1750 on a plantation in South Carolina, had little or no opportunities to own their own farms, move about freely from town to town, vote, hold office, have control over their own families, set up businesses, go to school, and so on. In other words, the social structure of slavery did not allow African Americans to have control over their own lives. Rather, their lives were largely, if not totally, controlled by the social structure of slavery. The social structure of slavery is a good example of what Durkheim (1966) was talking about when he said that social phenomena are external to us and yet coercive on us; that is, many times they influence us much more than we realize.

With the great prejudice and discrimination toward African Americans via slavery for 200 years (from the 1660s through the 1860s) and great prejudice and discrimination via segregation for another 100 years (from the 1860s through the 1960s), African Americans could not get ahead, could not have choices, could not be individuals, and could not be responsible for their own destinies because their destiny was already largely, if not totally, determined by the existing external and coercive social structure.[1]

The example of slavery and African Americans is just one example of a social structure giving little or no opportunities to a group of people. Another example is the placing of Native Americans on reservations, where they were restricted in where they could live and how they could make a living. For example, many Native American nations were hunters and gatherers, agriculturalists, or some combination of the two.[2] In a number of instances, however, they were moved from areas where there were plenty of animals to hunt, berries to pick, and fertile land to farm to areas that had few animals, few berries, and less fertile land. As a result of this forced migration, they were placed in a physical environment that provided them with less means of survival.[3]

If we want to decrease the kind of crime that places people in social structures of little or no opportunities for them to survive and sustain themselves, we will need to give people opportunities, specifically *legal* opportunities. Otherwise, just to survive, people may, for example, rob other people, burglarize houses, steal cars, sell illegal drugs, engage in prostitution, become pimps, or use violence to protect their territory in a neighborhood to continue to make money selling drugs. These are the kinds of crimes that are more likely to occur when people do not have legal opportunities to get ahead.

A good theoretical analysis of these types of crimes can be found in Merton's (1968) article, "Social Structure and Anomie" (recall our theoretical discussion in Chapter 1).[4] In a nutshell, Merton stated that when people have legitimate opportunities to get ahead, most will use these legal channels to get ahead. However, if we create social structures that do not allow people to get ahead, such as the social structure of racial/ethnic and ethnic prejudice and discrimination, the social structure of poverty where people live in neighborhoods of high unemployment and hence have few legitimate means to get ahead, and the social structure of deplorable schools that greatly limit what people can do, would like to do, and are capable of doing, we should predict more crime—more stealing (cars and other goods), burglaries, robberies, selling of drugs, prostitution, and so on.

On the other hand, if we give people more legitimate opportunities, they are less likely to resort to illegitimate means to get ahead and hence we should see a decrease in crimes related to little or no opportunities. These thoughts raise the following question: How can we give people more legitimate opportunities? Answering this question leads us to what we can do to solve or greatly decrease the kinds of crimes based on little or no opportunities.

What Can We Do?

There are a number of actions we can take to help us decrease crime because we will help people have more legitimate opportunities to get ahead. In the following paragraphs, let us discuss what we can do.

Public Education and Opportunities

One step that we can take to give people more legal opportunities is to give them an excellent public education from Grade 1 through Grade 12. In Chapter 7 on unequal public education, I mentioned that we could have a much better public education system by hiring only certified teachers, hiring more teachers so that class sizes are smaller (especially for low-income children who need extra help), creating smaller schools so that a higher percentage of the student body can take part in the extracurricular activities that go on at school and hence feel like a part of the school community, making sure that we have well-maintained schools, and making sure that the teachers and students have the equipment they need to do a good job such as having enough chalk, pencils, paper, textbooks, microscopes, computers, and band instruments. As young Americans get to attend good-quality public schools and get a firm grounding during the first 12 years of their educational careers, they will be more prepared to pursue legal opportunities and less likely to resort to illegal opportunities, thereby reducing crime.

More Jobs and Opportunities

If people have more legal opportunities, they will be prepared to take on more legal jobs. It follows that we will need to provide more legal jobs for people. As I mentioned in Chapter 4 on poverty, during the first few years of the 21st century, the United States has had an overall unemployment rate of between 5% and 6%. People living in low-income areas of our country (e.g., inner-city neighborhoods, rural areas, reservations) typically face even higher rates of unemployment. We need to find ways to create more jobs, especially more decent-paying jobs. This leads to the following crucial question: How do we do that?

There are a number of things we can do to connect people to legitimate jobs. One action we can take is to provide child care subsidies for poor people, especially single mothers, so that they can work outside the home. Without child care subsidies, many mothers either will not get these jobs or will not make enough money, especially at minimum wage and lower wage jobs, to be able to keep these jobs. The child care subsidy allows single mothers to get legitimate jobs rather than seek out illegitimate activities such as engaging in prostitution, selling drugs, and shoplifting. In other words, if we want to decrease crime, we need to give single mothers legal options that are better than illegal options.

Another step we can take via jobs and opportunities is to raise the minimum wage so that legal jobs have more of an enticement than do illegal activities. If people, after working 40 hours per week, do not have enough money to pay the rent, pay for food, and pay for other necessary expenses and they are still thousands of dollars below the poverty line, they will have greater incentives to consider illegal options as a way to survive and get ahead. In addition to increasing the minimum wage, we can increase the earned income tax credit to make up for the gap between the new minimum wage and the poverty line. Moreover, we can decrease income taxes and social security taxes on the poor. As we carry out these wage and tax policies, people who have more limited economic opportunities will have more money in their hands and will therefore have less incentive to commit crime.[5]

Taxes, Corporations, Jobs, and Opportunities

Another action that our country can take is to decrease taxes on corporations in exchange for creating more jobs in the United States. A drop in corporate taxes will create an incentive for corporations to stay in the United States and provide more jobs for Americans. As this happens, there will be more legal opportunities for people and less likelihood of people choosing illegal opportunities, thus acting to decrease crime.

Unemployment Compensation, Job Training, Job Placement, and Crime

We can also increase unemployment compensation for people who have lost their jobs so that they are less likely to resort to crime to survive while out of work. We can also provide for better job training and retraining so that people who have lost their jobs can find new legal avenues of employment and feel hope in being able to go this route rather than the illegal crime route. Furthermore, we can get better at placing people in jobs that open up so that they believe they have a good chance to get a legal job, again increasing the incentive to do legal work while decreasing the incentive to resort to illegal activities. As we increase our unemployment compensation, our job training programs, and our job placement programs, we will develop a social structure that allows Americans to choose legal opportunities over illegal opportunities, thereby decreasing crime.

Remember, there is no guarantee that every adult desiring a job will be able to have a job in a capitalistic economy. Even during good years, there is usually some unemployment. And during other years, there can be substantial unemployment. For example, during the mid-to-late 1990s, we had fairly low unemployment—in the 3% to 4% range (University of Texas, n.d.). However, during the early years of the 21st century, we have had higher unemployment—in the 5% to 6% range. Consequently, our unemployment compensation system is extremely important for providing people with a temporary cushion to help them in their transition from one legitimate job to another legitimate job.

Skilled Trades and Opportunities

Another action we could take in our country to decrease crime is to place more emphasis on skilled trades in our public high schools. Besides providing better quality public schools in general, which will help all students prepare for better jobs, providing better trade schools in particular, as one part of the improvement in the public school system, will give a number of high school students who do not want to go to college, or who are not capable of going to college, an avenue that will provide them with a skill that could give them good-paying legal jobs.

Decreasing Inequality and Increasing the Standard of Living

We can also take steps to decrease the overall inequality in our society so that more poor and near-poor people have a decent standard of living. So long as many Americans live considerably below the poverty line and many others live near the poverty line, there is an incentive for these Americans to find illegal means to survive. If they can have a decent standard of living via a legitimate job that pays above the poverty line, they will have less incentive to commit crime. So, it is in our vested interests as a country that we attempt to decrease inequality by redistributing resources somewhat if we want to have less crime (recall our discussion in Chapter 3 on what we could do to decrease inequality in our country).[6]

It is also in the vested interests of nonpoor Americans to want less inequality in that these people will be less likely to be the victims of crimes such as burglary, robbery, and car theft. So, the society will be safer in addition to having less crime if we create a more equal society where the needs of poor and near-poor Americans are met. As I suggested in Chapter 3 on the problem of rising inequality, we could create a more progressive income tax and social security tax and we could decrease the sales tax, meaning that poor and near-poor people would pay less tax and hence have more money left over from their paychecks to meet their family needs. Having more money left over each week after less tax has been taken from their paychecks will reduce the incentive to commit crime to survive.

Health Care and Less Crime

Also, having a national health care system could provide health care for many poor and lower income people who do not currently have any health care (for more details, see Chapter 10 on health care). During recent years, more than 40 million people in our country have had no health care coverage (Pear, 1999). Having a health care system would decrease inequality and would therefore provide a base for a decent standard of living, thereby decreasing the incentive for people to undertake illegal activities to survive.

The past few paragraphs have suggested that the way we tax ourselves and the kinds and amounts of social services we provide can lift up poor and near-poor people in our country and in turn create disincentives for them to commit crime, thereby decreasing crime. Although it might seem unrelated at first, on further reflection, we can begin to realize that the way we tax people and the amount and kind of social services we provide people are key factors that can decrease crime.

As you can see, I have focused mainly on what we, as a society, can do to help people get legal jobs and remain working at legal jobs. That is, we create a social structure that gives people more legal opportunities to have a decent standard of living. The more we can restructure society to give people more legal opportunities in the form of skills, education, and decent-paying jobs as well as more income via lower taxes and more and different kinds of services that provide a base for all Americans to live decently, the less incentive they will have to commit crime.

Registering of Guns and Crime

Another factor that we might consider is the registering and licensing of all guns in the United States. Although this action does not relate to the main thrust of the preceding paragraphs that emphasized giving people more legal opportunities, it may help to reduce crime in two ways. One way is by decreasing homicides in the United States. The total number of firearm homicides in 1995 was 15,835 in the United States versus 34 in Japan (Booth, 2004, p. 199). That is, although our U.S. population is only twice the size of Japan's population (Population Reference Bureau, 2002), the number of U.S. firearm homicides was 466 times that of Japan.

Besides decreasing the homicide rate in the United States, the registering and licensing of all guns in the country could help in not giving as easy access to firearms to people who intend to commit crime. This is not foolproof; people can still get firearms in illegal ways. However, we can make it more difficult for these people to get firearms. Less access should help in decreasing crime.

"Front-End" Investment and Crime

I have focused mainly on what our society can do so that people will be less likely to commit crime in the first place. In other words, if we invest a lot into preventing crime, our investment should eventually pay off. If I am right, we would not need to invest so heavily in other things (e.g., police, courts, prisons) after people have already committed crime. For example, once people have committed crime, we need more police to catch suspects; more lawyers, judges, and courts to try and convict defendants; and more prisons and prison personnel to hold the convicted criminals. If we, as a society, could invest more in the "front end" of this process so that people have less incentive to commit crime in the first place, we could spend less on the "back end" of this process, that is, the police, courts, and prisons. No doubt, there will still be a need for police, courts, and prisons, but the need should be less as we invest in restructuring society to give people more legal opportunities.

Job Training in Prisons and Job Placement

Speaking of investment in the back end of this process, another action we could take is to provide more and better job training in prisons and more job placement for men and women who have finished their prison terms. Many men and women who have completed their time in prison are let out of prison with nowhere to go and no job to help them stay legal. If our prisons could train people to become carpenters, electricians, plumbers, machinists, repair technicians (e.g., cars, appliances, televisions, computers), and other kinds of skilled and semiskilled workers, men and women leaving the prison system would have a better chance to find legal jobs and hence more incentive to "go straight," that is, to remain legal. Moreover, if the prison system had an effective job placement system, men and women leaving the prison system would have a much better chance of getting legal jobs and "going straight" in their lives. With such a combination of job training and job placement programs in our prison system, we could take another step toward decreasing crime.

The job training and job placement programs both are a part of the larger idea of rehabilitating people in prisons rather than punishing them only. In the past in the United States, our prisons have been mainly a place of punishment with little or no rehabilitation. If we want to decrease crime, especially the recidivism among people who previously committed crimes and served time in prison, it makes sense to increase rehabilitation, especially in the form of providing job training and job placement. In other words, if we can rehabilitate people leaving prisons by giving them job skills and then placing them in jobs, former prisoners will have the opportunity to make a decent living in a legal manner, thereby decreasing crime.

Moreover, if people leaving prisons have gone through job training and job placement for months or even years, they have had a chance to think about the possibility and to anticipate and plan for the day when they will be free and have legal and decent-paying jobs. This set of social conditions will help them to go through what sociologists call *anticipatory socialization*, where they will have thought about what it will be like to be free and working in legal, decent-paying jobs. Once they are working at their new jobs, they will be members of new reference groups (recall our discussion in Chapter 1) that should help them to get connected with other people who are working at legal jobs. It would seem, therefore, that providing job training, job placement, and new reference groups will work to prevent former prisoners from returning to crime, thereby decreasing crime.

Ideally, if we could get these people into an entirely new social environment, where the effects of differential association (recall our theoretical discussion in Chapter 1) with former friends and acquaintances who had helped them to commit acts of crime in the past were no longer a problem, the chances for recidivism and hence crime would diminish. Coming out of prison, having a new job skill, being placed in a decent-paying job, being a member of a new reference group at work, and being placed in a new social environment separated from their former reference group that promoted and reinforced acts of crime—altogether, these new social conditions should act, at the back end of the process to reduce recidivism, thereby decreasing crime.

To get former prisoners to move to another geographic location where they no longer interact with their former colleagues in crime, we might need to provide them with some kind of incentive such as a certain amount of money per month for a while. These people, with their new decent-paying jobs and new reference groups (and, it is hoped, new friends and legal role models), will have established a new legal way of living and can carve out a new legal way of life.

The policies I have suggested, both front end and back end, will require investment of money and expertise. To the degree that the actions we take in our society at the front end, where we provide better education, training, child care subsidies, more decent-paying jobs, a higher minimum wage, and lower taxes on poor and near-poor people, are combined with the actions we take at the back end, where we provide job training, job placement with new and legal reference groups and new and legal role models, and new geographic locations away from old illegal reference groups and role models, all of these front-end and back-end policies taken together should help us to substantially decrease crime in our country.

More Emphasis on the Front End

As you can see, my way of decreasing crime is a little different from others in that I put much more emphasis on changing the front end of the process by providing a lot more legal opportunities for people so that they have the incentive to do legal work and remain legal. The more we can invest in these front-end policies, the less we will need to invest in the back-end policies.

Other people will emphasize adding more police in total, more patrol cars, more police walking neighborhood beats, more surveillance cameras, more harsh punishment, and more prisons. We can go this way and put the money we invest into decreasing crime in doing these kinds of things. I am sure that these kinds of things will reduce crime somewhat because they will increase the certainty that someone committing a crime will get caught and be punished more severely. Research suggests that certainty of punishment adds to the deterrence of crimes (Paternoster, 1989). This is all well and good, and I think that these kinds of actions will be effective to a degree.

However, given the limited resources our society has to work with, investing in expanding the legal opportunities for Americans not only will address what causes crime but also will create the type of social structure that will reduce crime year after year and decade after decade.

A Dilemma: Capitalism and Crime

A dilemma that we, as Americans, need to face and deal with in some way is the dilemma of capitalism and its relation to the emphasis on materialism and consumerism in our society and crime. As Merton (1968) noted, one of the main reasons why people break the norms of the society to gain money and material things is the emphasis in our society on seeking these kinds of things. Merton noted that in our society, these are the things "worth striving for" (p. 187). He added,

> The cultural emphasis placed upon certain goals varies independently of the degree of emphasis upon institutionalized means. There may develop a very heavy, at times a virtually exclusive, stress upon the value of particular goals, involving comparatively little concern with the institutionally prescribed means of striving toward these goals. (p. 187)

So, when "there is an exceptionally strong emphasis upon specific goals" (Merton, 1968, p. 188), such as we have in our society on making money and having many material things (e.g., homes, things in our homes, cars), we should predict that a number of people will use illegal means to get these things.

In the society in which we currently live, we turn on the television and see advertisements every few minutes urging us to buy certain products or services. Hence, during a typical 2 hours of watching television, we will be bombarded with 40 to 50 commercials telling us to buy this or that. We pick up a typical magazine and see many advertisements in it as we thumb through the pages. We read a newspaper and see as many as five advertisements on one page. We drive along highways and see numerous signs telling us to buy all kinds of products and services. We use a computer and see ads among our e-mails or when we use the Internet. We attend sporting events and see ads nearly everywhere we look in the stadium, park, field, or fieldhouse. Ads are even fixed on restroom walls for us to look at and reflect on as we take care of our biological needs! As you can see, we are surrounded in our culture with the emphasis on consuming (Ritzer, 2005).

It will therefore not be surprising to you how a certain amount of crime is related to the emphasis on consuming. Why is there such an emphasis? The nature of capitalism helps us to understand why. As you know, capitalism is based on owners of the means of production producing products and services and selling these products and services to make a profit. Consequently, the more products and services that can be sold, the more profit that can be made. This is where advertising and commercials come into the picture. Advertising and commercials try to get people to want to buy all kinds of products and services.

So, back to the dilemma that I mentioned a few moments ago. On the one hand, most of us in the society want to decrease crime. Yet on the other hand, as Merton (1968) pointed out, if "there is an exceptionally strong emphasis upon specific goals without a corresponding emphasis upon institutional procedures" (p. 188), people are more likely to commit crime. So, what do we do in a society that emphasizes the goal of obtaining many material things and yet wants to decrease crime?

Do we decrease the cultural emphasis on the goals, that is, decrease the emphasis on obtaining material things? We could do that by banning many kinds of advertising and try to emphasize a less materialistic lifestyle. However, it is hard to imagine that we could do this in our society for a number of reasons. First, our current citizens are accustomed to having a materialistic lifestyle. It is hard to imagine our giving up such a lifestyle. Second, it is hard to imagine in capitalism that corporations would accept not being able to advertise their products and services. If this scenario is accurate, for the foreseeable future we will need to work at decreasing crime within a social structure that has, as Merton put it, an extreme emphasis on having a lot of things.

Given such a set of social conditions, it appears that we will need to work to institutionalize ways for people to have access to the material things via legal means. This brings us back to our discussion earlier in this chapter regarding the need to provide more legal opportunities for people— providing better public schools, teaching people trades so that they can make good incomes, giving people a good education so that they can have decent-paying jobs, and so on. So, even though it is not realistic, for the time being, to decrease the emphasis on obtaining material things as a way of decreasing crime, it does seem realistic to increase the legal opportunities for people as a way of decreasing crime.

Questions for Discussion

1. In addition to the points discussed in this chapter, what else can we do to decrease crime in the United States?

2. What should we do in our country? What is your reasoning?

3. What do you predict will be done in our country during the next 5, 10, and 15 years? Why?

4. Should we hire more police and build more prisons? Why?

5. Is it possible to decrease the emphasis on material goods and consumerism in our country? What is your reasoning?

6. If you were poor, why might you commit crime?

7. If you were rich, why might you commit crime?

8. What one thing could be done in society to greatly decrease crime?

9. If there are not enough jobs for people and we want to decrease crime, what can we do?

10. If you were stealing cars, selling drugs, or working as a prostitute, what do you think would stimulate you to have a legal job?

9

How Can We Solve the Problem of Drugs?

The drug problem is a difficult one in that there are opposite views in our country on what we should do about it. Many Americans believe that the current policy of keeping certain drugs illegal (e.g., marijuana, cocaine, heroin, methamphetamines) is the best policy even though a number of users still manage to get and take these drugs and others still manage to sell them at a profit. So, even though the current policy is far from foolproof, many Americans believe that keeping certain drugs illegal is still the best policy.

There are, however, other Americans who are of the opinion that our current drug policy is morally wrong because it does not allow adults (defined as people who are at least 18 years old and therefore can vote, serve in the military [and possibly die in wars], marry, have children, have jobs, and live independent lives) the freedom to decide for themselves whether or not they want to take a certain drug. These Americans believe that adults should have the freedom to decide for themselves what drugs they use. In other words, they believe in giving Americans as much freedom as possible vis-à-vis the choice of drugs. So, in this social problem of drugs, a key issue is deciding how much freedom adults should have. Currently, there is much disagreement over this issue, making the area of drugs a social problem.

Because Americans disagree, it is not surprising that our current drug policy is inconsistent. For example, nicotine, the drug in cigarettes, cigars, and pipe tobacco, is legal, yet it kills more than 320,000 Americans per year (Nadelmann, 1988, p. 24). Alcohol, the drug in beer, wine, and whiskey, is legal and kills 80,000 to 100,000 people per year (p. 24). Yet illegal drugs such as marijuana, cocaine, and heroin kill less than 4,000 people per year (p. 24).[1] So, the drugs that kill many people are legal, whereas the drugs that kill few people are illegal.

What should we, as a country, do about the drug problem? Should we continue the current policy of keeping a number of drugs illegal even though we know that a lot of people still use and sell these drugs and we know that we spend a lot of money trying to stop people from taking these drugs and trying to stop people from bringing these drugs into our country? Should we keep the drugs nicotine and alcohol legal even though they cause many more deaths than all illegal drugs combined? Should we continue to be inconsistent with our drug policy, where some drugs are legal and even highly promoted (e.g., beer ads on television) and other drugs are kept illegal? Should we legalize the illegal drugs so that we are consistent in our drug policy? Or should we make alcohol and nicotine illegal? That is, should we be consistent in our drug policy, or should that not matter? Should we decide for adults what drugs they can use? Or should we let adults decide? These are all important questions that our country has yet to answer, thereby producing this social problem.

So, what can we say to help us solve this social problem? Let us consider the advantages and disadvantages of going one of three ways and see what you think: (a) keep the current policy of keeping alcohol and nicotine legal while keeping marijuana, cocaine, heroin, amphetamines, and other drugs illegal; (b) change the current policy to legalize drugs for people who are 18 years old or over; or (c) take a step-by-step approach by legalizing one drug at a time to see what happens, for example, first legalize marijuana, see what happens, and then go from there.

Keep the Current Policy

Advantages

One advantage to keeping the current policy is that by keeping marijuana, cocaine, heroin, amphetamines, and other drugs illegal and harder to get than they would be if they were made legal, the fact that these drugs are defined by the general society as illegal and bad will no doubt deter a number of people from using them. A number of children will be taught, "These drugs are bad for you, so don't use them." This admonition from parents, teachers, other adult leaders, and fellow playmates will send a strong message to many children that will deter them from using illegal drugs. To the degree that this admonition works, these young people will not use these drugs and hence will not be in the situation of becoming addicted and having problems with these drugs.

A second advantage of keeping such drugs illegal is that the mere difficulty of obtaining the drugs and their cost, along with the unknowns of what is actually in the drugs one is buying, will deter a number of people from ever trying and using these drugs and hence prevent them from having drug-related problems in their lives, for example, in their jobs and with their families.

Disadvantages

There are a number of disadvantages to continuing to carry out the current policy of keeping a number of drugs illegal. For example, we are beginning to see a pattern of people robbing drugstores not for the money but rather for the drugs kept at the drugstores. Some people will go to extremes to get their drug of choice to use or sell. This means that pharmacists and others who work in drugstores will be working in an atmosphere that is less safe. Also, potential robberies of drugstores for the drugs will cause drugstores to need to spend more money for security precautions and to pay more for insurance policies insuring the products and people in the drugstores, hence increasing the cost of doing business.

Another problem with the current policy is that we still have millions of Americans using illegal drugs. Currie (1993) reported that in 1991, 20 million Americans used marijuana, 6 million used cocaine, 5.7 million used hallucinogens or inhalants, and 700,000 used heroin (p. 66), so that a total of 32 million Americans used some type of illegal drug. Yet there were only 234,600 prisoners in our state and federal prisons who were serving time for drug-related crimes (Myers, 2001, p. 237). In other words, less than 1% of illegal users ended up in jail. Nadelmann (1988) stated, "Those agencies charged with drug interdiction, from the Coast Guard and U.S. Customs Services to the U.S. military, know that they will never succeed in capturing more than a small percentage of the illicit drugs being smuggled into the United States" (p. 6).

Another problem with the current policy is that because a number of drugs are illegal, their illegal status promotes a lot of drug-related crime such as violence, gang warfare, and organized crime. Individuals, gangs, and criminal organizations sell illegal drugs to make a profit. Becker (2001) noted that the world market value of illegal drugs is "at several hundred billions of dollars—in the same league as the markets for cigarettes and alcohol" (p. 32). In other words, selling illegal drugs is a profitable business. To protect their profit, sellers will resort to violence to keep rival sellers out of their market.[2] Hence, there are threats, fights, shootings, stabbings, and murders to protect the business. Consequently, there will be more illegal business and more violence as a result of our keeping drugs illegal.

Another problem with the current policy of keeping such drugs illegal is that it costs billions of dollars in taxes to enforce. It costs American taxpayers $40 billion per year to try to stop the flow of illegal drugs into the country and to have the criminal justice system try, convict, and jail illegal users and sellers (Brownstein, 1992, p. 220). By jailing these illegal drug users and sellers, more than 60% of all federal prisoners and 22% of all state prisoners are in prison for drug offenses (Myers, 2001, p. 237). Taken together, 26% of all state and federal prisoners in our country are in jail for drug-related offenses (p. 237). Each new prison cell costs $80,000 to build, and each new prisoner costs the state or federal government approximately $25,000 per year to house (Eitzen & Zinn, 2003, p. 290). Brownstein (1992) asserted that our current drug policy and its high costs divert our scarce resources from addressing other social problems such as poverty, health care, education, and more job training and employment opportunities (p. 231). Moreover, Eitzen and Zinn (2003) noted, "Despite all of our efforts, we have not stopped the supply; we have only dented it and made the drugs that enter the United States more expensive" (p. 390).

Our current drug policy also causes our government to be involved in the domestic politics of other countries in terms of trying to get these other countries to destroy their coca and opium crops. Our government's action raises the ethical question of how much right or authority we have in destroying farmers' crops in other countries. Although our policy may decrease the quantity of drugs smuggled into our country, it does not endear us to these countries and their farmers because we want them to destroy one of their major means of economic survival.

Another disadvantage to keeping drugs illegal is that it takes more police time to try to stop users and sellers of these drugs. Nadelmann (1988) reported, "State and local police were estimated to have devoted 18 percent of their total investigative resources . . . to drug-enforcement activities" (p. 15). By spending nearly one fifth of their time on trying to catch people using and selling illegal drugs, police have less time to focus on many other crimes such as murder, robbery, rape, burglary, and corporate crime (e.g., tax evasion, false advertising, stealing retirement funds from employees). In other words, our current drug policy is a major influence on how police allocate their time and hence on what kinds of crimes they focus on more and what kinds of crimes they focus on less.

Another disadvantage to keeping a number of drugs illegal is that as alleged sellers and users are arrested and brought to trial, the result is an increase in time and expense in our court system. For example, existing judges need to spend more time on drug-related trials, or more judges are needed to take care of the many drug-related trials. Also, more courtrooms, court personnel, and prosecuting attorneys are needed, resulting in more expense for the taxpayers who pay for the new courtrooms and salaries of these additional governmental officials. In a nutshell, more court time and taxpayer expense go to address one kind of crime—drug crime.

A consequence of keeping drugs illegal is that they cost more to buy (Becker, 2001, p. 32). Yet as Becker (2001) noted, "The fact remains that most illegal drugs remain popular and available, regardless of price" (p. 32). For people who do not have the money to buy these drugs, they are more likely to commit other kinds of crime such as shoplifting, burglary, robbery, and prostitution to get the money to buy the drugs. So, although we do not intend for our current drug policy to cause an increase in other kinds of crime, our policy makes these drugs cost more, and this in turn causes users who do not have enough money to buy these drugs to commit more crime to be able to purchase the drugs. Hence, our current policy unintentionally acts as a latent dysfunction for those who are the victims of other kinds of crime (recall our discussion in Chapter 1 on functional theory and the idea of latent dysfunctions).

Another major disadvantage of the current policy of keeping certain drugs illegal is that the users do not know what they are getting (Nadelmann, 1988, pp. 20–21). That is, when they buy drugs off the street, they do not know how pure the drugs are and what else is contained in the drugs. So, the users are taking a chance when they use the drugs, not knowing how the contents of their purchases will affect them.

Another disadvantage of the current policy is that users can get labeled as *deviants, addicts, criminals,* and so on, and these labels can work to keep them from getting legal jobs. They can be stigmatized and hence have a harder time in living and working in the legal world. This labeling process acts on them as a latent dysfunction for their daily survival.

Finally, another disadvantage of the current policy is that it does not give adults freedom to decide for themselves whether or not they want to consume certain drugs. This is especially upsetting to people who have a libertarian philosophy and believe that "drug use is a personal matter in which the state has no business" (Fine & Shulman, 2003, p. 60). Szasz (1972) asserted, "It is none of the government's business what idea a man puts into his mind; likewise it should be none of the government's business what drug he puts into his body" (p. 75).

In our country, we believe in giving people a lot of freedom. However, in the area of drug use, we have decided that adults should not have this freedom. They can use nicotine and alcohol, but they cannot use marijuana, cocaine, and heroin. This raises the following question: Should adults be free to decide what drugs to use, or should the government decide via legislation?

Legalize Drugs for Adults

As in the current policy of keeping certain drugs illegal, there are also advantages and disadvantages to legalizing drugs. Let us consider these advantages and disadvantages.

Advantages

If we legalize drugs such as marijuana, cocaine, heroin, and amphetamines, as well as other drugs that are now illegal, one advantage is that we should observe less crime. As these drugs can be obtained legally, say in drugstores, the drugs will be cheaper in cost because the supply will be greater; for example, farmers in other countries and our country will not have their crops burned, and there will be no stopping of drugs at our country's borders. For those people who have lower incomes, they will not need to commit as much crime in terms of stealing, burglary, robbery, prostitution, and so on to buy the drugs. Consequently, in addition to drug crime decreasing, other kinds of crime will also decrease.[3] Also, other kinds of crime, such as gang violence and gang warfare, could decrease because there will not be as much territorial fighting and violence to keep other gangs out of a certain neighborhood given that there will not be drugs to sell.[4] People instead will be able to go to a local drugstore to purchase drugs at a lower cost, thereby knocking out, or at least seriously damaging, the drug market by gangs. Finally, we should see a decline in organized crime because the legal selling of drugs will hurt the market for drugs through organized crime networks.[5] In fact, it might not be in the vested interests of gangs and organized crime to see drugs legalized because customers will be able to buy them in a legal and less costly way and will know the purity and contents of the drugs they are purchasing, thus resulting in a substantial decline in the purchase of drugs through gangs and organized crime. Hence, the legalization of drugs could be dysfunctional for the survival of gangs and organized crime. However, it is likely that most Americans would like to see the decline in gangs and organized crime.

With a decline in various kinds of crime, with a decline in gang violence and gang warfare, and with a decline in organized crime, we will not need to spend as much of our tax money on trying to stop crime. There will not be as many police officers needed because buying and using drugs will no longer be crimes. There will not need to be as many prosecuting attorneys, judges, and courtrooms, and this will in turn mean less taxpayer money going to these things. Also, there will not need to be as many prisons and prison personnel because there will not be people being convicted of buying and selling illegal drugs because the drugs are now legal to buy (recall that approximately 26% of all state and federal prisoners are in prison for drug-related crimes) (Myers, 2001, p. 237). So, our taxes could go down given the decrease in these costs. Or we could take this unspent tax money and use it to address other social problems.

Legalizing drugs will allow police to spend more of their time on other kinds of crime, and this in turn will help us to deter these crimes. Police will be able to devote more of their time and attention to crimes such as burglary, robbery, rape, murder, and corporate crime. Police will also not need to be stretched so thin in terms of needing to develop so many skills and so much knowledge about so many different areas of crime, for example, in the area of drugs. The more police can specialize in fewer areas of crime, the more effective they can be at catching criminals and decreasing crime.

Also, there could be less police corruption. Police officers do not make high salaries. They can therefore be tempted to supplement their incomes by receiving payoffs from drug dealers. Nadelmann (1988) noted,

> What makes drug enforcement especially vulnerable to corruption are the tremendous amounts of money involved in the business. Today, many law enforcement officials believe that police corruption is more pervasive than at any time since Prohibition. Repealing the drug prohibition laws would dramatically reduce police corruption. (pp. 19–20)

Another advantage to legalizing drugs is that it gives adults in our country more freedom. Being able to choose whether or not to take drugs and what drugs to take will give adults more freedom.[6] They can choose, rather than the government deciding for them, whether or not to use marijuana, cocaine, heroin, and other drugs. As we mentioned in the previous section, it may be fruitful for our country to have a national discussion on whether or not adults should have freedom to choose when it comes to drugs.

Another advantage of legalizing drugs is that there will be less labeling of the users of drugs. Right now, people who use various illegal drugs are many times labeled as *deviant*. This label makes it harder for these people to live normal lives. If drugs are legalized, the stigma of a drug user could lessen and hence the effects of this stigma could be less in drug users' lives.

Users would be less likely to be discriminated against as much and thus would be more likely to be able to get and keep legal jobs, pay their bills, and interact with people who have legal jobs, thereby making users more likely to engage in legitimate activities in various areas of their lives and less likely to commit crime.

Another major advantage of legalizing drugs is that the government will be able to tax the sale of these drugs.[7] This could be a major new source of revenue for local, state, and federal governments. Given that all levels of government need more revenue, taxing the sale of marijuana, cocaine, heroin, and other drugs could prove to be beneficial for the government and, at the same time, take the pressure off of taxpayers who pay other kinds of taxes.

Another key advantage of legalizing drugs is that there will be new legal jobs. For example, people growing plants in other countries that produce marijuana, cocaine, and heroin will not be stigmatized and will not have their governments working against them to grow these crops. This will in turn bring in more revenue for these countries and their governments in addition to the farmers and middle people who process the crops into the drugs and transport them to countries that consume the drugs. Legalization will also create new agricultural jobs in our country in that farmers will be able to grow plants that produce marijuana, cocaine, and heroin. We will also need more people in our country to process the plants, check for purity, transport the drugs, and sell the drugs at local drugstores. So, legalizing these drugs would create more jobs in our country and especially would give a stimulus to people who work in agriculture and related jobs.

Finally, users will know the purity of the drug and what else is contained in the products they purchase. Hence, legalizing drugs will protect the health of the users more than the current policy does.

As you can see, there are many advantages to legalizing drugs—probably more than most Americans realize. Now let us consider the disadvantages to legalizing drugs.

Disadvantages

One of the main disadvantages is that we will probably see a certain amount of drug addiction increase if we legalize drugs such as marijuana, cocaine, and heroin. This is understandable given that these drugs will be more accessible in drugstores and will be lower in cost. As a result, people will be able to buy these drugs more easily as they do now with alcohol and nicotine.

Before we accept the idea that there will be an increase in the use and addiction of these new legal drugs, let us make a few cautionary statements.

Nadelmann (1988) noted that decriminalization of marijuana in a dozen states during the 1970s "did not lead to increases in marijuana consumption" (p. 29). Also, The Netherlands decriminalized marijuana and saw its consumption actually decrease. Also, research indicates that during the late 19th century when there were no drug laws, drug use was roughly the same as it is today with all of our drug laws and regulations (p. 29). So, although there may be an increase in the use and abuse of these newly legalized drugs because they are cheaper to get and easier to buy, this is not a certainty.

If there is more addiction, we, as a society, will need to be ready for this. This would mean, for example, that we will need more drug rehabilitation facilities and personnel than we have now. As a consequence, some of our tax money that we receive on the sale of these drugs and that we save in needing fewer police, judges, prosecutors, trials, and prison cells will need to be set aside to pay for additional facilities and personnel who will help those users who abuse drugs.

With this increase in addiction, there will probably be some increase in people being absent from their jobs, doing lower quality work, and having family problems. We should predict and expect these kinds of things as people have the choice of taking these drugs. With problems of more absenteeism, lower quality work, and family problems, we will need to have counseling and rehabilitation facilities to address these consequences, as with the expected increase in addiction.

So, although we should see a decrease in time required of, and expenses incurred by, police, courts, and prisons because the criminal justice system will not be involved in trying to catch, try, convict, and jail people who use illegal drugs, there will possibly still be other expenses. Which way will be more expensive? No one knows for sure, but I would predict that it will be much lower in cost to provide rehabilitation centers and counseling personnel than to have police spend 18% of their time investigating drug-related activities, have the court systems devote court time to drug-related crimes, use tax revenues to pay for prison space for 26% of all state and federal prisoners imprisoned for drug-related offenses, and have federal law enforcement spend billions of dollars in trying to shut down drug operations in other countries in an attempt to slow down the flow of illegal drugs into our country. All in all, I predict that we would find a huge difference between the expense of keeping certain drugs illegal and making them legal.

Finally, even if we legalize drugs, I predict that nicotine and alcohol will continue to be the two main drugs of choice by most Americans and therefore will continue to be the two most abused drugs in our society. The reason I say this is that Americans are already accustomed to using these two drugs at parties, at bars, and in their homes. These two drugs have been an accepted part of our culture for more than 75 years. Also, there are powerful corporations that have strong vested interests in continuing to make much profit by keeping these two drugs as the drugs of choice for Americans. If I am right, the continued use of alcohol and nicotine would tend to temper the increase in the use of these other drugs. However, if corporations can advertise these new legal drugs on television, on the Internet, on billboards along highways, and in newspapers and magazines, the use of these new legal drugs could rise sharply. The increase in use would therefore depend on the regulations we put on these drugs, for example, whether or not they can be advertised.

Take a Step-by-Step Approach by Legalizing One Drug at a Time

There is a third approach that could be labeled as a "middle-of-the-road" approach. As a compromise between the two preceding policies of keeping certain drugs illegal and legalizing drugs, we could experiment with one drug at a time where we legalize it, sell it in drugstores, see how it affects our society, and then assess where we want to go next.[8] For example, we could legalize marijuana and see what happens in our society.[9] Does it hurt our society too much or not very much? On balance, is it better to legalize it and deal with the problems this causes, or is it better to keep it illegal and deal with the problems this causes? We might not know the answers to these questions unless we experiment by legalizing marijuana and then see which way society would rather have it—legal or illegal.

I suggest that we experiment first with marijuana, rather than cocaine or heroin, because the public sees marijuana as the least harmful in terms of being not so powerful and not so harmfully addicting. If it were legalized, it could be regulated like alcohol and nicotine, where adults would be allowed to buy it at drugstores. Let us discuss the advantages and disadvantages of using this middle-of-the-road approach, or step-by-step approach, by using the example of legalizing marijuana.

Advantages

In terms of advantages, we could tax the legal sale of marijuana, and this would help our society to address our government debts and deficits. We also know that we will be able to tell users what the marijuana contains and how pure it is. That way, people will know what they are getting and how it will affect them.

Another advantage will be that our prison population will decrease to the degree that it is made up of people convicted of selling or using marijuana. This decrease in the prison population will take the pressure off of our crowded prisons and decrease the need to build costly new prisons.[10] Also, we will not need to use police and court time in trying to catch and try alleged sellers and users. All of these things—less use of police time, less use of court time, and a decline in the prison population—will mean that we will not need as much tax money for these kinds of things.

Another advantage will be to give more freedom to adults who feel that they should have the right to decide whether or not to use marijuana. For those Americans who have a more libertarian philosophy, they will appreciate being able to have this additional freedom in their lives.

Even though marijuana is illegal, it is one of the largest cash crops in our country (Fine & Shulman, 2003, p. 58). For example, Nadelmann (1988) noted that marijuana production in our country is believed to be "a multi-billion-dollar industry" (p. 9). If it were legalized, it could help a number of farmers in our country to have another legal crop with which to make a profit and survive in the profession of agriculture. Also, we could export it, thereby helping us with our balance of trade deficit. People growing it, processing it, and selling it would mean that more jobs would be created, thereby helping the economy.

Disadvantages

There are also disadvantages to legalizing marijuana. More people could get addicted to marijuana because it would be more easily available to those who want to use it. If this is true, we, as a society, would need to spend more tax and insurance money to rehabilitate people who are addicted.

When people drive under the influence of marijuana, their reaction time slows down and they are more likely to make misjudgments and cause more car accidents, as when they drive under the influence of alcohol. We always hope that people under the influence of a drug will use good judgment and not drive. Many people use good judgment. However, we all know that some people do not use good judgment, and some of these people end up causing car accidents that can be tragic for themselves and others—as we have seen over the years with drinking alcohol and driving. Here the problem is not so much the use of the drug in and of itself as it is the use of the drug in a situation that could cause harm to themselves and other people. Consequently, if marijuana were legalized, we should predict that there will be at least some rise in car accidents and in the more material costs that accompany these accidents.

Where to Go From Here?

So, what should we do in our society with respect to illegal drugs? As you know, I, as a sociologist, cannot answer a "should" question. Sociology cannot tell the society what it should do. I, as a sociologist, have pointed out major advantages and disadvantages to keeping drugs illegal, to legalizing drugs, and to taking a step-by-step approach to legalizing drugs.

There are many opinions as to what we should do. With other social problems, most people want to decrease poverty, decrease crime, decrease various forms of prejudice and discrimination, and decrease pollution. As for drugs, however, there is, at this time in our country, no clear direction as to which way we want to go.

To make matters more cloudy, this issue can also become a hot political issue to use against one's political opponent by trying to get voters to be emotionally against one's opponent rather than to have voters calmly and objectively look at the pros and cons of taking each approach. As a result, candidates will fear raising this issue in a campaign. Consequently, we have not had much of an open and objective discussion on this issue in our country.

As to where we are and where we need to go in our country, it seems to me that it could be fruitful for our country to have an honest, open, and objective[11] national discussion and lay out the advantages and disadvantages of going each way. We, in sociology, along with other disciplines of psychology, political science, economics, anthropology, history, biology, and chemistry, could add what empirical evidence we have about drugs and their effects on the body, drugs in history, or drugs in other cultures so that we could have the most informed discussion possible.

Given such a national discussion, we could grow in our awareness, knowledge, and understanding of how drugs fit within the larger social structure of our society.[12] Also, with such a discussion, we would be in a better situation to make a decision as to which direction we want to go in our country.

Questions for Further Discussion

1. Are we spending too much money trying to stop people from using or selling illegal drugs and trying to stop others from bringing illegal drugs into our country? If so, what should we do?

2. Should we be consistent in our drug policy, that is, either make alcohol and nicotine illegal or make marijuana, cocaine, heroin, and amphetamines legal? What is your reasoning?

3. Should adults have the freedom to use the drug they would like to use? What is your reasoning?

4. If we legalized drugs that are currently illegal, what kinds of laws would we need to create?

5. If marijuana were legalized, what would be the consequences in our country?

6. Should marijuana be legalized? What is your reasoning?

7. Should we try an experiment with marijuana and legalize it in certain cities or geographic areas and then see how it works and what the consequences are?

8. Should we try a step-by-step approach to legalizing drugs, that is, legalize one drug and see whether that goes okay and then legalize another drug and see whether that goes okay? What is your reasoning?

9. Should we legalize all illegal drugs? What is your reasoning?

10. What do you predict will occur with respect to the legalization of illegal drugs in your lifetime? What is your reasoning?

10

How Can We Solve the Problem of Health Care?

Health care is an immensely difficult problem in our society. Every industrialized country except the United States has national health care for all of its citizens (Kerbo, 2006, p. 37). There are currently 46 million American citizens who have no health care coverage, that is, roughly 1 of every 6.5 Americans.[1] So, this means that when these people are sick or have a toothache, it will cost them a lot more money than it will cost those of us who have health care coverage. As a result, they will be more likely to put off going to the doctor or dentist and to not get any medical help at all.[2] Employers that help to pay health care costs are finding that they are being put in a less competitive position because of rising health care costs (Durbin, 2005; Maynard, 2006). Employees are having a harder time paying for their health care. Many retired people living on moderate to lower incomes are needing to use more and more of their retirement incomes to pay for prescription drugs (Carroll, 2003). Poor people who are on Medicaid (the government health care program for the poor) still might not receive health care because some doctors will not accept them due to the compensation being too low to cover their expenses (Yetter, 2005). In short, we have a number of problems with our current health care system, and so far our country does not know which way to go or what to do.

Consequences of Our Problem

To this point, our country has resisted going to a national health care system, with the result that millions of Americans go without any health insurance and hence put off going to doctors and dentists until their situations are so serious that they cannot put them off any longer. The consequence of this for these individuals is that they are in much worse health, they endure a lot of needless pain and discomfort, and "18,000 adults die each year because they lack health insurance" (Unger, 2006, p. A10). Adults miss work days and lose pay, and their employers lose efficiency and profit. Students lose school days, are in danger of getting lower grades, and cannot participate in school activities. A sad example of this is that in the state of Kentucky alone, "338,000 Kentucky children covered by Medicaid never saw a dentist last year" (Yetter, 2005, pp. A1–A2). There are a number of reasons why these children did not receive dental care. First, more than half of the dentists in Kentucky will not treat people who are on Medicaid because the dentists do not get reimbursed enough from Medicaid to cover their expenses (pp. A1–A2). Second, because parents on Medicaid are poor, many times they do not have means of transportation to get to dentists who will treat their children (p. A2). Third, many parents work at jobs where it is difficult to get time off from work, or they are not permitted to leave work, to take their children to dentists (p. A2). Fourth, even if parents are allowed to take time off from work, many times they lose pay that they cannot afford to lose, especially when they make such low wages (p. A2). Added to all of these negative consequences on poor people is the probability that there will be further cuts in the Medicaid program, where poor people could be charged premiums and need to make copayments (Pear, 2006a). (A *copayment* is a certain amount of money a patient needs to pay the doctor each time the patient sees the doctor.) This cut would mean that 13 million poor people, or approximately one fifth of all Medicaid recipients, "would face new or higher copayments for medical services like doctor's visits and hospital care" (p. A3).

There are consequences for the elderly in our country. Under the current system, the elderly, especially those who have moderate incomes or who live in or near poverty, need to use what little social security income and retirement income they accumulated through their previous jobs to pay for their monthly prescription costs, meaning that they have less money to pay for food, rent, transportation, and other expenses (Carroll, 2003, p. A1). Due to the high cost of U.S. drugs, some Americans travel to Canada to get their drugs or buy drugs over Canadian Internet sites.[3] As you can see, there is a serious conflict of vested interests (see our theory and causal model in Chapter 1) between the elderly who need prescription drugs at lower prices because these prices take so much of their retirement incomes and the drug companies that argue they need to charge higher prices because of the cost of doing research on new drugs.[4]

There are consequences for many corporations given our current health care system. By paying for health coverage of their employees, corporations need to add on to the prices of their products, making their products less competitive in the marketplace than the products offered by corporations from other countries that do not need to pay for health insurance because their governments pay for such coverage.[5] For example, it is estimated that General Motors needs to add $1,500 to the price of each car to pay for its employees' health care coverage.[6] Also, American corporations are cutting back jobs. For example, General Motors announced in November 2005 that it was going to cut 30,000 jobs in North America by 2008 (Durbin, 2005; Maynard, 2006). One of the main reasons for cutting back these jobs is rapidly rising health care costs. The result is that American corporations are less competitive and American workers are losing their jobs, in part, due to health insurance costs. So, from various perspectives (e.g., the poor, children of the poor, the elderly with moderate to lower incomes, employees, corporations), there are many unfortunate consequences of our current health care system.

There is also a relationship between one's income, one's health, and one's health care coverage. Kerbo (2006) asserted that "good health is to some degree unequally distributed through the stratification system" because "adequate health care is unequally distributed" (p. 37).[7] Once again, we see how one social problem, such as substantial and rising inequality (see Chapter 3) or persistent poverty (see Chapter 4), is connected to another social problem, in this case inadequate health care. As we can find ways to decrease inequality and poverty, we can improve the health care of people and vice versa; that is, once we improve the health care of people, our inequality and poverty can decrease. Once again, we can see how the solving or ameliorating of one social problem can help to solve or ameliorate another social problem.

A final major consequence of the current health care system that all Americans face is the constant pressure of rising health care costs. This affects everyone, although it affects disproportionately more Americans with less income. For example, from 2000 through 2005, yearly health care premiums increased at least 2.6 times the yearly inflation rate and increased as much as 8 times the inflation rate (see Table 10.1). (A *premium* is the amount of money a person pays each month for health care coverage.)

Table 10.1   Inflation, Premiums, and Times the Premiums Are More Than
Inflation Rates

Year	Inflation Rate (%)	Yearly Premium (%)	Times the Premiums Are More Than Inflation Rates
2000	3.1	8.2	2.6
2001	3.3	10.9	3.3
2002	1.6	12.9	8.1
2003	2.2	13.9	6.3
2004	2.3	11.2	4.9
2005	3.5	9.2	2.6[a]

a. See Unger, L. (2006, January 30). Support swells for universal health care. *Louisville Courier-Journal*, pp. A1, A10.

Why Don't We Have a National Health Care System Like All Other Industrial Nations?

There are a number of reasons why we do not have a national health care system. The main reasons seem to boil down to (a) the vested interests of maintaining the current health system, (b) the desire to maintain political power, (c) the ideology of individualism that we have created in our country (again, refer back to our theory and causal model of conflict and social change in Chapter 1), (d) the fact that many Americans currently have access to excellent health care, and (e) the fact that most Americans do not know much about possible alternatives to our current system.

Vested Interests

Doctors, dentists, drug companies, and health insurance companies want to make money. The current health care system results in doctors and dentists making high incomes and drug companies and health insurance companies making much profit. In other words, as Daniels, Light, and Caplan (1996) put it, "The failure of national health care lies in the fact that comprehensive reform threatens powerful, wealthy interests" (p. 17).[8]

Let us take the example of drug companies versus the elderly and the issue of the price of prescription drugs. Due to the high cost of drugs produced in the United States, more and more Americans have been buying their drugs from Canadian companies. Moreover, a national poll found that 70% of respondents agreed that "it should be legal for Americans to buy prescription drugs outside the United States" (Lester, 2003, p. A5). During recent years, U.S. drug companies have been fighting to keep the importation of prescription drugs from other countries illegal, arguing that they need to recover the costs of doing research to create new drugs and that drugs from other countries could be "unsafe" (Carroll, 2003, p. A7). The "unsafe" argument is probably not a valid one given that "Canadian authorities subject all drugs sold in the country, including those made in the United States, to testing similar to that conducted by the FDA [U.S. Food and Drug Administration]" (p. A7). As you might expect, there is disagreement over whether and how much money drug companies need to do research on new drugs and how much profit they make. The drug companies say that the average cost to do research on a drug is $800 million (p. A7). Hence, they say they need the higher prices on their drugs to make up for expensive research. Anne Northup, a Republican congresswoman from Kentucky who supports the importation of prescription drugs to decrease drug costs for Americans, said that, on the contrary, drug companies are making 18% profit after paying their taxes—"a breathtaking amount of profit" (p. A7). From a profit point of view, why would drug companies want to change the current health care system if they are making that much money?[9]

Assuming that this situation is a key barrier to any change in the health care system, we will need to address this impediment in some kind of equitable and fair way, that is, lower costs for drugs but reasonable compensation to drug companies. In addition to profit concerns of drug companies, the income concerns of doctors under a possible national health care system would also need to be addressed. For example, evidence from a study conducted on Canadian physicians found that when these physicians are paid well, they are less resistant to a national health care system (Globerman, 1990, pp. 22–23). Reaching an equitable and fair situation will, no doubt, require considerable national discussion in our country.

Political Power

Fear of losing one's political power is another barrier to creating a national health care system. Presidents and members of Congress attained their positions of power in part, and possibly in large part, through the contributions of drug and insurance companies. The more this is the case, the less these holders of political power want to "bite the hand that feeds them," that is, the less they want to risk losing contributions and risk losing their political offices and the power that goes with them.

People who want to be in Congress or be president, or people who want to remain in Congress or win a second term as president, need lots of money to get elected and to remain in power. As Weber (1968) and Marx and Engels (1992) noted more than 100 to 150 years ago, money and power tend to be closely related. In the cases of Congress and the president (as well as state offices such as the governor, state senator, and state representative offices), substantial money is needed to gain political power. Some of the people and organizations with a lot of money in our country are doctors, dentists, drug companies, and health insurance companies. Hence, these individuals and organizations can have a big influence on who gets to have political power in our country. Moreover, they can influence legislation that can be favorable to their vested interests. As a result, money, political power, and the kinds of laws and social policies that we have can be, and many times are, closely tied to each other.

Ideology

With respect to the third reason of why we do not have a national health care system, ideology, we need to refer back to the beginning of our country. Historically, our early American ancestors did not want monarchies to rule us, they did not want governments to grow too big and intrude in our lives, and they did not want a feudal caste system where people could not move up and down the social class system (although our country did institutionalize such a caste system for African Americans in the form of slavery and later in the form of legalized segregation and discrimination, and we did accept many traditions of Europe where males ruled and did not allow women to be upwardly mobile). Our ancestors began to construct a new social reality[10] where we tended to emphasize more individualism, getting ahead, having the chance to be upwardly mobile, being more industrialized, being more of an urban nation, and being a more capitalistic nation. Such a view of life has meant that we have been slower to accept government programs and services than have people in European countries. This belief system, where we rely less on the government, has been a barrier to our having a national health care system.

Excellent Care

A fourth reason why we do not have a national health care system is that many Americans currently have excellent health care coverage. If you think about it for a moment, many of us have nice homes and cars, we live in nice neighborhoods, we send our children to good schools, we take nice vacations, we can pay our bills, we have excellent food and plenty of it, we have all kinds of options for recreation and leisure activities, we can go to shopping malls that are filled with all kinds of things to buy, we can purchase new technological innovations that are constantly coming onto the consumer market, and so on. In other words, many Americans are living a very good lifestyle. In addition, many of us already have excellent health care provided by our employers and our own monthly payments. Given this enviable situation, why would we be concerned about health care? Like other personal problems, if we are not bothered by them, we tend to take less interest in them—whether the problem is poverty, racial/gender inequality, or lack of health care.

Lack of Knowledge

Finally, a fifth reason why we do not have a national health care system is a lack of knowledge. That is, we Americans do not have much, if any, knowledge about how health care systems work in other industrialized countries. Other than hearing a little about Canada having lower prices on prescription drugs than the United States, we know little or nothing about other health care systems with which to discuss options for our current system. Many times, when people do not know much about something, they fear the unknown. Instead of being able to seriously consider various health care options based on knowledge, so far we have based our decision making on knowing how we benefit from the current system but do not know how we could benefit from any other system. So, our philosophy has been, "It is better to stay with the known than to try the unknown."

If Americans can be given knowledge about elements of health care systems from other industrialized countries such as Germany, Canada, and Great Britain and then can discuss the pros and cons of these elements, Americans would have knowledge that could allay their fears about the unknown, and this in turn could help them to seriously consider other ways of having a health care system. Until we have this knowledge, we will be ambivalent about trying anything new and different and hence will conclude, "It is better to stay with the known."

In short, the vested interests of wanting to make much money, the desire to maintain political power, the ideology of individualism and believing that we should not rely on the government (or should rely on it as little as possible), the fact that many Americans already have excellent health care, and not knowing about other ways we could have a health care system together present a formidable barrier to our having a national health care system. It logically follows that if we can address these barriers, we may be able to consider a new kind of health care system that is a substantial improvement over the one we currently have.

Possible Social Conditions Leading
to a New Health Care System

Given that we have so many negative consequences with the current health care system—for the elderly, the poor, workers, and corporations—it is an appropriate time in our country to seriously discuss what we can do. In fact, a recent poll found that "nearly nine out of 10 Americans think the current system is broken" (Unger, 2006, p. A10). One immediate step we could take is for the president of the United States to appoint a bipartisan commission to study what could be done and make recommendations to the president and Congress. This commission would need to be made up of researchers, policymakers, and other experts in the field of health care and should not be made up of people based on their political leanings. Congress, likewise, could create a bipartisan committee of senators and members of the House of Representatives to study alternatives in depth. President Bill Clinton took the first serious step of looking into the health care problem in his first term in office during the early 1990s. Neither the country nor Congress, however, seemed to be ready to face the problem at that time, so nothing came of Clinton's efforts. More than 10 years have passed since then. We have increasingly become aware of the problem and the consequences that result from our current health care system. Problems of the elderly, the poor, families, released workers, and the competitiveness and profit of corporations—all related to our current health care system—are more and more in the news and talked about among people (Unger, 2006, pp. A1, A10).

More than ever, Americans are realizing the problem and concluding that something new and different needs to be done. For example, a national poll taken in October 2003 found that "54 percent of Americans are dissatisfied with the overall quality of health care in the United States" (Lester, 2003, p. A5) and that this was a 10% increase in dissatisfaction from 2000 to 2003. The poll also found that 62% of Americans said they "preferred a universal system that would provide coverage to everyone under a government program, as opposed to the current employer-based system" (p. A5). Moreover, 80% of those polled agreed that "it is more important to provide health care coverage for all Americans, even if it means higher taxes, than to hold down taxes but leave some people uncovered" (p. A5).[11] Even 49% of doctors favored creating a national health care system (Unger, 2006, p. A10). These recent opinions by the American people suggest that they are ready for new, bold, and more comprehensive initiatives.

The current president or the next president, through the appointing of a bipartisan commission, could take the initiative to begin a national study and discussion of this social problem. Because the problem of health care is negatively affecting so many people in so many ways, it seems appropriate to seriously address this problem with new, bold, and creative thinking and discussion.

No one knows what might be concluded by such a commission or congressional committee. The reason I say this is that even though there are many problems with the existing health care system and we, as Americans, are becoming increasingly aware of these problems, we still know there are barriers that will stand in the way of members of a commission or congressional committee tackling our health care problem in a new and creative way.

Even if a commission or congressional committee is created, it does not necessarily follow that this group (or these groups if both are created) will be creative and innovative and will make bold visionary recommendations. If anything, we would predict that such an entity would, in all probability, be more tepid and piecemeal in its recommendations because existing vested interests could cripple any boldness and creativity. For example, President George W. Bush promoted health savings accounts, where people can put money in a tax-free savings account (Howington, 2006). This money can be used to pay for medical expenses that people's regular high-deductible health care plans do not cover ("Health Savings Accounts," 2006). This could help some people who can afford this. But many lower income people who are not covered by Medicaid will not be able to afford this, resulting in their continuing to go without health care coverage. In other words, the suggested policy is not bold enough to help most of the 46 million people who currently do not have health care or most of the elderly who are poor or near poor and cannot afford such health savings accounts.

This being said, is there any ray of hope in this scenario? I see two scenarios that could offset what I just outlined. First, there does seem to be more and more discussion in newspapers and on television about our health care problem. More people are reading and thinking about this issue, and they are watching and listening to reports on this issue. This issue is not going away in the minds of the American people. This interest and concern could very well be the key that could eventually counterbalance, or even outweigh, the current barriers to creating a better health care system.

If the American people increasingly come to the conclusion that there are too many problems with our current health care system, they could put increasing pressure on the president and Congress to do something substantial about it. Given these pressures, the president and Congress could very well address our health care problem in a much more comprehensive way than has ever been done before.

The key to our taking bold and decisive steps with regard to health care is whether or not the American people get more and more concerned about this social problem.[12] If they do not get more concerned, the five barriers we discussed earlier in the chapter will probably win the day and there will be only a tinkering with the current health care system, with the result that the elderly, the poor, many workers, and many corporations will continue to face their respective problems that we discussed previously. If, on the other hand, the American people become increasingly concerned about this problem, we could see the creation of a new health care system that addresses the existing concerns and finally meets the needs of all the American people.

Criteria for a New Health Care System

No one knows what a new American health care system would look like exactly. Assuming that we create such a system, it will be unique. It could take on elements from other countries such as Germany, Canada, and Great Britain (more about these countries later). Because these nations also are democratic, are predominantly capitalistic, and have high standards of living, and because they already have national health care systems, we could eventually borrow various elements of their health care systems to create our own unique system.

Assuming that we create a new health care system some day, no one knows what this new system would look like, but I think we can create criteria that we might want this new system to satisfy and then follow with a discussion of the elements that could carry out these criteria:

1. The new health care system will be based on the health care needs of Americans instead of on their ability to pay for health care.[13]

2. All citizens will be covered by this new health care system.[14]

3. All citizens will receive good-quality care.

4. All citizens will receive care in a timely way.

5. As an integral part of the new health care system, there will be an emphasis on preventing illness by creating incentives and providing education for people to live a healthy day-to-day lifestyle, thereby decreasing the cost of the health care system.[15]

6. As an integral part of the new health care system, there will be an emphasis on preventing illness by providing periodic checkups for all Americans so that any potential illnesses can be caught early, thereby decreasing the cost of the health care system.

7. By providing a new health care system that allows Americans to live a healthier day-to-day lifestyle, this system will allow Americans to live more enjoyable day-to-day lives.

8. Doctors, dentists, nurses, and other health care personnel will be paid reasonably for their services.

9. Drug companies will be paid reasonably for their products.

10. Health insurance companies that could be devastated by the loss of business with a new health care system will be compensated sufficiently so that these companies can move into other areas of insurance or other areas of business.[16]

11. For-profit hospitals and hospital chains can choose to run their hospitals for a reasonable amount of compensation from the government or to sell their hospitals to the government for a reasonable amount of compensation.

Elements of a New Health Care System to Satisfy These Criteria

Given the preceding criteria of what we would want to have in a new health care system, let us now begin to piece together elements of a health care system that would satisfy these criteria.

Element 1: Buying Bulk and Competitive Bidding

One element of our current health care system that could be used nationwide is what the Department of Veterans Affairs (VA) is currently doing with millions of American veterans of military service (Pear & Bogdanich, 2003; Weigel, 2006). The VA saves a lot of money in buying its drugs for our veterans "through bulk purchasing arrangements—using generic drugs where possible—and competitive bidding" (Pear & Bogdanich, 2003, p. A7). That is, instead of veterans needing to buy drugs at high prices in drugstores, the VA has drug companies bid on which companies will provide particular drugs at the lowest prices. The VA buys these drugs at much lower prices and in turn provides these drugs to the veterans who "pay just $7 for up to a 30-day prescription" (p. A7). If we would use this same strategy for all Americans where the federal government bought drugs in bulk, used generic drugs where possible, and had the drug companies bid on drug contracts, the costs to the federal government for providing prescription drugs to Americans would be less, the taxes needed from the taxpayers would be less, and the costs to individual Americans needing the drugs would be less, yet the drug companies selling the drugs would still make a profit—although not as much profit as they do in the current health care system. By using this strategy, we could satisfy some of my criteria such as providing health care based on needs rather than ability to pay (Criterion 1), covering all Americans (Criterion 2), and allowing drug companies to earn a reasonable profit (Criterion 9).

Element 2: Social Conditions That Promote a Healthy Lifestyle

Another key element of a new health care system is to get people to live a healthier lifestyle. This could save the taxpayers of our country a lot of money in that there would not be as many expenses that need to be paid out by our new health care system.

One way to promote a healthier lifestyle is to give Americans tax breaks if they achieve certain goals throughout the year. For example, if, on examination, people are designated as being overweight, they can get an increasing tax break as they lose a certain amount of weight. The same principle could be applied to people who have high levels of cholesterol and blood pressure and other measurable indicators of health. Also, if we could invent some way of measuring the amount of healthy exercise people do over the course of a year, people could receive more of a tax break depending on their amounts of healthy exercise. Likewise, if we could measure and report the healthy intake of food (e.g., fruits, vegetables, grains) and the amount of rest needed for good health, these could also be factors going into decreasing people's taxes. Such a system not only would decrease our taxes directed toward health care but also would help Americans to live a more enjoyable lifestyle (Criterion 7).

A second way to promote a healthy lifestyle is to have early education and more education on how to live a healthy lifestyle. From the first grade onward, students can be taught about eating the right kinds of foods and getting sufficient exercise and rest. Part of a national health care plan would be to allocate more money to school systems to teach children about living a healthier lifestyle and to mandate that all public schools can provide only food that is part of a healthy diet and that any unhealthy food (e.g., soda, pizza, burgers, fries) cannot be provided at the schools.

The new health care system could provide for all Americans to have periodic exams (medical and dental). There are a number of reasons for these exams. First, any problems will be discovered early, meaning that there is a better chance to cure people. Second, the early detection and cure will mean that Americans will endure fewer catastrophic illnesses that are so costly, thereby decreasing the costs of our health care system. Third, the early detection and cure will mean that people will be in less pain for a shorter amount of time. This element of a new health care system will satisfy the following criteria: Criterion 2 (all Americans are covered), Criterion 3 (all receive good-quality care), Criterion 4 (all receive care in a timely way), Criterion 6 (decrease health care costs via periodic checkups), and Criterion 7 (people live a more enjoyable lifestyle).

Besides these direct measures that will help people to live a healthier lifestyle, an indirect measure that will help people to live healthier lives is decreasing economic inequality in our society (see Chapter 3 on how we can decrease inequality). Budrys (2003) noted that there is a consistent correlation between one's social class and one's health (pp. 211–213). She reported, "Accumulated evidence tells us that social class must be a more important factor than anyone previously realized" (p. 211). She added, "The association between social inequality and poor health is consistent and powerful" (p. 211). She therefore concluded that if we want to have better health in our country, we need to decrease inequality. She added that although we have tried to convince people to eat less fatty foods, quit smoking, and exercise more, "reducing economic inequality can be expected to have a much bigger impact on the health of the population than trying to convince individuals to change one at a time" (p. 214). Hence, as we work to decrease our social problem of inequality, we will also help to ameliorate the health problems we have in our country and, in so doing, to decrease our health care costs. As a consequence, we will address Criteria 5 and 6, which work to prevent illness, and Criterion 7, which results in Americans living a more enjoyable day-to-day lifestyle.

Element 3: A Progressive Federal Income Tax to Pay for the New Health Care System

To have a health care system that covers all Americans (Criterion 2) with good-quality care (Criterion 3) and in a timely way (Criterion 4), we will need enough money to pay for it. Because the new health care system could be one national system, probably the most realistic way of procuring the money in the United States would be to get the money at the federal level via a progressive income tax.

The additional money needed might not be that much more than we are currently spending for health care. Currently, we are paying taxes for Medicare to help retired people have health care and for Medicaid to help poor people have health care. Also, many employers are paying insurance companies to insure many of us who work. Finally, most of us who have health insurance plans with our employers also contribute from our salaries and wages to our health insurance plans. As we would change to a national health care plan, we could abolish Medicare, Medicaid, employer, and employee costs and instead pay one cost through our income taxes to the federal government.

Which system—our current system or a new health care system—will cost us more? I do not know. We would need to have experts consider all of the variables and analyze how the costs of the two systems compare. We could find that once we subtract (a) the costs of Medicare, (b) the costs of Medicaid, (c) the costs of more than 50 bureaucracies that carry out Medicare and Medicaid, (d) the costs to employers to insure their employees, (e) the costs to employers that lose business because of being less competitive in the marketplace, and (f) the costs to employees to help pay for their own health insurance, the difference between the costs of the current system and the costs of the new system might not be that much. Given that all other industrialized countries are able to pay for national health care systems for their people, it seems that we too could figure out a way to do likewise.

By having a progressive income tax that brings in sufficient revenue to pay for our new national health care system, we will be able to pay doctors, dentists, and other health care professionals reasonably (Criterion 8), pay drug companies reasonably (Criterion 9), and compensate health insurance companies and hospitals sufficiently (Criteria 10 and 11). Such a strategy would help to reduce the opposition of health care professionals, drug companies, and health insurance companies that have vested interests in wanting to continue with the current health care system.

A controversial matter that our country will need to address relative to living a healthy lifestyle is the influence of large and powerful corporations that advertise on television, in newspapers and magazines, and on the Internet for people to eat fatty foods, such as cheeseburgers, fries, and pizza, as well as to drink alcohol and, until recently, to smoke cigarettes. These companies are businesses that want to make a profit. The problem is that what they sell works to the detriment of people's health. We will need to address this issue as we discuss how we can live a healthier lifestyle.

Element 4: One Health Care System

In Canada, each province and territory creates its own health care system so long as it meets minimum standards set by the federal government (Health Canada, n.d. b). Both the federal and provincial governments contribute to paying for their health care system with both federal and provincial taxes. We, in the United States, could go in the same direction, but because our states are strapped for raising tax revenue, we might choose to have the federal government pay for the health care system and for there to be one system that is standard throughout the country. A big advantage of this kind of system is that when people move to other areas of the country, as many Americans do, they do not need to go through the process of dropping one system and signing up for another system, they do not need to fulfill any waiting period for being on the new system,[17] and they know what the specific benefits are because these benefits do not change from state to state. If we went in the direction of having one national system, we would know that all citizens would be covered all of the time (Criterion 2).

Element 5: Medical Care Geographically Dispersed

The new health care system would need to make sure that all Americans can get to medical care within a reasonable amount of time. That is, is there reasonable access to medical care in rural areas of our society, in inner cities, and on reservations? If not, our new health care system would need to be planned accordingly to address Criterion 3 (good-quality care) and Criterion 4 (receive care in a timely way). The problem of geographic access is a bigger problem for poor people because they are less likely to have their own means of transportation[18] to get to medical care. To provide access to all Americans, we might need to give doctors, dentists, and nurses certain financial incentives to induce them to live in geographic areas that provide more access to many Americans.

Element 6: Public Education—Teaching Diet, Cooking, and Exercise

The public school system, from kindergarten through the 12th grade, can be given sufficient funding to teach students (a) what food to eat and not to eat, (b) how to cook nutritious food to make it taste delicious, and (c) how to get healthy exercise.

To make all of this come about, a new major in college, called *healthy lifestyles,* can be created. A student who majors in healthy lifestyles will learn details on how to have the best diets possible, how to cook the most nutritious meals possible, and how to get the healthiest exercise possible. Once students in this major learn these three areas, they learn how to teach these three areas of healthy living to others, especially elementary, middle, and high school students. Public schools will incorporate these new kinds of teachers and courses into their classrooms. Also, corporations can be given certain tax breaks if they hire these people with a healthy lifestyles major to teach courses to employees and provide the appropriate facilities needed.

As a part of this element, public school systems throughout the country could be required to serve nutritious food only and could be prohibited from having on the school grounds food that has high saturated fat, high cholesterol, and/or high sugar content.[19] Foods such as pizza, burgers, fries, cake, cookies, and soft drinks could not be served. Diets at the schools could be made up of foods such as fresh fruit and vegetables; baked and grilled chicken, turkey, and fish; and water, low-fat milk, and tea.

Carrying out this element will especially address Criterion 5 (providing education for people to live a healthy day-to-day lifestyle) and Criterion 7 (living a healthier lifestyle that leads to more enjoyable day-to-day lives). By having such an educational system in place, we will prevent many serious illnesses from occurring (Criterion 5) and will cut the costs of our new national health care system (Criterion 6). By taking such an approach to health care, we will take a more proactive way of dealing with people's health rather than a reactive way where people do many things to hurt their health, get ill, and then treat their illnesses with expensive operations and/or costly pills that people take for the rest of their lives.

Element 7: Try Out the New Health Care System in a Select Number of States

Before we switch the entire country to this system, we could select a few states, which would like to be a part of this initial stage, to carry out the new system. By doing this, we could benefit in two ways. First, we could "work out the bugs" before we went nationwide with the new health care system. Second, we could show the rest of the nation that the system could work and how it could work. Canada unintentionally did something like this during the late 1940s and early 1950s when four provinces began health care plans, thereby demonstrating to the rest of the Canadian population that, yes, such a system could work in Canada (Graig, 1999, p. 125). If a few states showed the American people that a national health care system could work and how it could work, Americans who were dubious might be more inclined to give such a system a try.

Before we go to the next section of this chapter, I want to make you aware of some historical points. First, as far back as 1916, Congressman Meyer London of New York called for a national health care system of some sort (Walker, 1969, p. 299). During the few years before his call, labor unions, state legislatures, the American Medical Association, the American Pharmaceutical Association, and the American Hospital Association voiced concerns over the lack of health care in our country and called for and carried out various studies. At the time, there were a number of state legislatures that were seriously considering creating state health care plans. So, consideration of some form of a more comprehensive health care system has been around in our country for nearly 100 years.

Advantages of a National Health Care System

There will be a number of advantages if and when we have some form of a national health care system that meets the preceding criteria. One advantage is that many poor people who were on Medicaid but now are on the new health care system can get good-quality care in a timely fashion. As we noted previously, a number of doctors will not treat poor people on Medicaid because they do not get paid enough from Medicaid to cover their expenses. Likewise, hospitals do not always get reimbursed for the services they provide because the people who are served cannot pay for the medical services or are on Medicaid, and Medicaid does not always reimburse hospitals enough to cover their costs (Daniels et al., 1996, p. 5). Our moving to a national health care system will give good-quality health care in a timely fashion to poor people and to currently uninsured people and will, at the same time, reimburse doctors and hospitals sufficiently so that they can cover their expenses.

The working poor, who have no coverage through their employers but who are not poor enough to be on Medicaid, will benefit greatly from a national health care system. This would mean, for example, that the 46 million people in the United States who currently have no health coverage will get coverage (Unger, 2006, p. A1). Because they will have access to good-quality care in a timely fashion, they not only will live healthier lives but also will experience less absenteeism from work, make 10% to 30% more income per year, and thereby increase tax revenues to various levels of government (Budrys, 2003, p. 233). So, having a healthy workforce will work to the vested interests of corporate profits and government revenues.

The elderly, by having access to a national health care system, will immediately have a higher standard of living because they will not need to use portions of their social security and retirement incomes to pay for prescription drugs. Hence, they will not need to choose between paying for their pills and paying for their food.

In addition to the preceding groups who will benefit, corporations and workers in our country will benefit greatly. Corporations will benefit because they will not need to pay for health coverage for their employees, and this will immediately make American businesses more competitive in the world marketplace. For example, General Motors recently reported that it needs to add $1,500 to the cost of each car it sells to pay for the health care coverage of its current and former employees.[20] Workers will benefit in that because there is less cost to the company for having workers, corporations can hire more workers or at least retain the ones they have.

Another key benefit that would come from having a national health care system is that there could be just one bureaucracy administering this system versus the many bureaucracies administering the current health care system. Currently, we have one bureaucracy administering the Medicare system for the elderly and 50 state bureaucracies administering the Medicaid system for the poor.[21] If we had a national health care system administered by one bureaucracy, we could get rid of 50 bureaucracies with all kinds of different rules and benefits and replace this system with one bureaucracy with one set of rules and benefits for all Americans. It would seem that one bureaucracy focusing on one health care system would be more efficient and less costly than many bureaucracies. Moreover, although libertarians wish to have hardly any government (except for the defense of the country)[22] and conservatives do not want big government with many bureaucracies (Fine & Shulman, 2003, pp. 9–11), it would seem that given the alternatives of the current system with many bureaucracies and a new system of one bureaucracy, and given the realities of the modern complex society in which we live, a number of libertarians and many conservatives would opt for the new system that has one bureaucracy.

A related advantage to having one government bureaucracy is the fact that doctors and dentists, under the new national health care program, will not have "the huge cost and time burden of processing insurance forms and negotiating with insurers regarding clinical decisions for their patients" (Daniels et al., 1996, p. 5). Doctors and dentists will now deal with one bureaucracy rather than more than 50 government bureaucracies and many insurance bureaucracies. This should greatly relieve the headaches of office personnel in doctors' and dentists' offices of needing to work with all of these bureaucratic entities.

Another advantage of having a national health care system is that Americans, wherever they move in the United States and whether or not they have a job, will always be covered. In the current system, as Americans move from one job to another job, they may or may not have health insurance. Also, in the current system, as Americans get laid off from their jobs or get downsized, they lose their health coverage, putting all members of their families at risk. With a national health care system, everyone will be covered no matter what happens in their job situations or where they move.

People will also know what their benefits are. Their benefits will not change from job to job. Such a situation will take away much stress that occurs now when people lose their jobs or move to other parts of the country. They will no longer need to worry about the health care part of their lives. It will be both constant and consistent.

Another advantage of having a national health care system has to do with the current situation of health maintenance organizations (HMOs) that provide health care for people. HMOs charge a flat fee to those who join. The people joining HMOs then know that they do not need to pay any more money to receive health care. That sounds good for these people. There are, however, two problems with this system. HMOs, knowing that they will get a flat fee, will want to have only healthy people in their plans so that the health care costs will remain low and the HMOs will make more profit. They will not want to sign up people who have serious illnesses that will cost a huge amount of money to treat, thereby decreasing the profit the HMOs make.[23] So, seriously ill people can be turned down for coverage by HMOs.

A second problem with the HMO system is that HMOs are not required to disclose the treatments they offer and the conditions under which they are offered, consequently making it next to impossible for people to know whether or not they will receive coverage. A national health care plan would include all people (Criterion 2), no matter how seriously ill they are, and would spell out for all Americans specifically what kind of care they will receive (Criteria 3 and 4).

Another major advantage of a national health care system will be to those people who currently do not have health insurance coverage. Under the existing health care system, every time uninsured people have something wrong with them, they take a huge risk in going to a doctor or hospital to seek care given the possibility that they will be charged thousands or hundreds of thousands of dollars that could devastate them for the rest of their lives. This situation is one of the worst consequences of our current health care system. With a national health care system, this tragedy for a number of Americans will be no more.

Disadvantages and Costs of a National Health Care System

One disadvantage of the new national health care system would be to executives of companies. In many situations, executives of companies currently pay the same premiums for health care coverage as do the employees. Although the premiums can take a considerable chunk of employees' wages and salaries, it takes only a small fraction from the salaries of executives. In essence, the current system of paying monthly premiums is regressive on employees. That is, they pay a higher proportion of their incomes on health care coverage than do executives. In this way, this system works to the advantage of executives. Recall that whereas worker pay to top corporate executive pay was a ratio of 40 to 1 in 1990, this ratio had increased to 419 to 1 by 1998 (Kerbo, 2006, p. 23). With executive pay increasing disproportionately to worker pay, this means that the monthly premiums that workers pay are more regressive than ever.

Another disadvantage will be to the drug and insurance companies. Given that the federal government will use competitive bids and bulk purchasing to see which companies can offer the lowest prices on drugs and services with insurance companies, these companies will not make as much profit as they are currently making. Consequently, their executive salaries and employee wages will not be as much as they are now. These companies will still make reasonable profits and compensation, but not what they currently make. So, these companies might not make the 18% profits that have been recently reported. And as you might predict, this is where there will be attempts by drug and insurance companies to discredit a new national health care plan. Given that the drug and insurance companies have "deep pockets" with which to mount a negative campaign against a national health care system, we, as sociologists, should predict that there will be advertisements on television and in newspapers and magazines vilifying a new system and that there will be members of Congress, who receive significant amounts of campaign contributions from these companies, waxing indignant about the perils of going down the path toward a "socialistic system."

Consequently, we, as sociologists, should predict that drug and insurance companies will succeed at slowing down the progress toward a national health care system. Whether their efforts result in halting or even abandoning the movement toward a national health care system, or whether it merely slows down the process, we should predict that a major barrier to getting a national health care system will be the actions of the drug and insurance companies—unless they can be assured of reasonable profits and compensation with the new system.

What Other Countries Include in Their National Health Care Systems

Before we conclude this chapter, let us see what other industrialized countries include in their health care systems so that we can get an idea of what we could do.

Germany

Germany was the first national health care system and was created during the late 1800s by Otto Von Bismarck. Graig (1999) noted that the German health care system "is among the most comprehensive in the world" (p. 50). The German system covers the following areas:

> Medical, dental, in-patient hospital care, prescription drugs, preventive care, and even rehabilitative treatments at health spas are covered. Patients do not pay deductibles, though there are minimal copayments for eyeglasses, dentures, prescription drugs, and the first fourteen days of a hospital stay. . . . Another measure of the breadth of German national health insurance is in the area of income replacement. Generous maternity benefits provide full pay during the period six weeks before birth through eight weeks after birth. (p. 50)

People below a certain income must be a part of this system (this includes approximately 75% of the population), and those above this income can choose to be a part of this system (Graig, 1999, pp. 50–51). Two thirds of the people who are above this income level, and who have a choice to be or not be a part of the national health care program, choose to be a part of it (p. 51).

To pay for the German health care system, employers and employees pay an equivalent amount into the system based on the employees' incomes (Graig, 1999, p. 52). The German system provides "the same benefits to the unemployed as to the employed" (p. 53). Unlike the current U.S. system, if German workers lose their jobs, they are still covered (p. 53). Patients can choose the doctors they want (p. 58). Doctors get paid based on negotiations between doctors' associations and the associations that run the national health care system.[24]

Canada

The idea of a national health care system in Canada began to take hold during the 1940s when a Gallup poll found that 80% of Canadians favored a national health care system and that the Canadian Medical Association, representing the doctors in Canada, also supported the creation of such a system (Graig, 1999, pp. 124–125). Canada's health care system is actually 12 systems representing 10 provinces and 2 territories (p. 123). Doctors have private independent practices and are paid fees for their services, but these fees have already been negotiated. Hospitals are mainly private non-profit organizations (p. 123). Each provincial and territorial health care system needs to meet four criteria set down by the federal government. First, each plan is comprehensive so that many services are offered. Second, each plan is administered publicly, that is, by an agency of the provincial government. Third, each plan is available to all people in the province. Fourth, people from one province, when visiting another province, can get medical service (called *portability*) (p. 126). In the process, a province can even buy out private health insurance buildings, computers, and employees (p. 126). Hospitals have remained private nonprofit organizations, but instead of being reimbursed by an insurance company, they are reimbursed by the government. Consequently, Canadian citizens, instead of paying doctors and hospitals directly or paying insurance companies to pay the doctors and hospitals, pay taxes to the government, which pays the doctors and hospitals (p. 127). Graig described the system this way:

> Preventing a two-tier system of health care made up of those who could pay and those who could not was a high priority. The federal government determined that the best way to avoid such a two-tier system was to take the dramatic step of outlawing private insurance coverage for any services covered under the provincial plans. Provincial universal hospital and medical plans thus took the place of all the various forms of insurance—private, not-for-profit, and public—that had existed up to that point. The provincial governments became the single purchasers of publicly insured hospital and medical care services. . . . By the mid-1970's, 95 percent of all costs of hospital and medical care services were paid through the provincial health plans and 0 percent of the population had private comprehensive health insurance coverage. (p. 128)

Graig (1999) also pointed out that Canadians go to the doctors of their choice and present a health card. As such, they do not need to fill out any forms, pay any copayments, or pay a deductible (p. 131). (A *deductible* is a certain amount of money that the patient needs to pay for health care first before the rest is paid for by the patient's insurance.)

As you might predict, there has been, and continues to be, conflict between the doctors' associations and the provincial governments as to how much doctors will be paid for the services they render. Provincial governments try to hold down rising medical costs, whereas doctors want to get a fair return on the services they provide. Hence, the negotiations are constantly difficult. Some doctors—approximately 1%—have left Canada as a result of these contentious disagreements (Graig, 1999, p. 144). As you can see, even with a national health care system, there will be built-in conflicts of interest. So, even if we go to a national health care system in the United States and achieve the goal of meeting everyone's medical needs, our system, like other national health care systems in other countries, will be in a constant process of conflict, negotiation, and change. Also, as you can see, these are the kinds of conflicts that will need to be hammered out if we go toward a national health care system.

Great Britain

Great Britain has what is called the National Health Service (NHS), which "the entire British population depends on" (Graig, 1999, p. 151) for its health care needs. The central government plays the key role, more so than in the United States, Germany, or Canada. That is, the government finances and runs the health care system (p. 153). The British health care system "provides essentially cradle-to-grave care, regardless of one's ability to pay, and is free to the patient at the point of service" (p. 154). (The phrase "free to the patient" means that patients do not pay money when they go to the doctor, but they pay the federal government in taxes and the government in turn pays doctors and hospitals.) In 1997, the government spent 6.7% of gross domestic product (GDP) on health care, whereas the United States spent 13.6% of GDP (or roughly twice as much money) on health care, yet various health indicators, such as infant mortality and life expectancy rates, were similar in the two countries.

So, although the British health care system costs less than half what the American system costs, two key problems that the British system has but that the American system does not have are long waiting lists to get medical service and shortages in technology (Graig, 1999, p. 154). On the other hand, although one of six Americans does not have health care coverage, all citizens of Great Britain are covered. Even prescription drugs (which have been a political "hot potato" in the United States) are nearly free in Great Britain. That is, some people need to give copayments, but 80% of the prescribed drugs are dispensed free of charge (p. 157).

As you might predict for any national health care system in any country, there will be conflicts of vested interests, constant desire for more or less of something depending on the interest groups, and so on. Great Britain is no different. Although Great Britain has been able to keep its health care costs relatively low, the British people want more services (Graig, 1999, p. 159). Also, the British population, like the American population, is aging, and hence there will be more elderly people requiring more services. Also, although the waiting time to see a doctor or get medical service has not increased in Great Britain since the 1960s, the number of people on waiting lists has increased (p. 159). So, an aging population, a demand for more services, more people on waiting lists, and an ever-changing economy make for constant change within the British health care system. Vested interests of patients, taxpayers, doctors, hospitals, and the government all work to put pressures on any national health care system. So, we in the United States, like the people in Great Britain and other countries, will need to change our health care system as different variables change.

The Future: What Could Happen?

No one knows what will happen in the near future.[25] However, I do have some ideas given recent trends in our country. The recent trends are that health care costs have increased substantially, making more people aware of this growing problem. Another trend is that more corporations and non-profit organizations are decreasing or eliminating coverage for their employees. For example, 69% of American corporations provided health care coverage for their employees in 2000, whereas only 60% provided coverage in 2006. One implication of this decrease in coverage is that members of the middle class, not just the elderly and the poor, are threatened with a decrease or outright loss of health care coverage (Beauchamp, 1996, p. 12). Such a trend, in all likelihood, will get the influential middle class more and more concerned about the situation of health care in our country. Another trend is that prescription drug costs continue to climb at such a rate that it is causing the elderly to use more and more of their social security and retirement incomes to buy the prescription drugs they need. This decreases their standard of living to the point where they have more financial stress during their retirement years. Another trend is that the number of uninsured in our country has increased from 44 million in 1999 to 46 million as of early 2006.[26] Another trend is that Medicare for the elderly and Medicaid for poor people are being cut back, making it more difficult for the elderly and poor people to get health care.[27] Another trend is that the health care problem is continually in the news these days, with the result that Americans are more aware of this problem and want something done about it. A front-page article titled "Support Swells for Universal Health Care" in the *Louisville Courier-Journal* on January 30, 2006, is such an example (Unger, 2006, pp. A1, A10). An editorial titled "Ailing Health System Needs More Than a Few Band-Aids" in the *Indianapolis Star* on February 21, 2006, is another example (Feldman, 2006, p. A6).

Given these trends, it appears that we are going in the direction of focusing more and more on this problem. This leads me to predict that unless something unforeseen happens, such as another war or natural catastrophe that would divert our attention and resources, we, as Americans, will increasingly demand that something more comprehensive be done. This increasing demand by many American people will finally get the wheels of the national government to turn in the direction of creating a new health care system that will include all Americans.

Questions for Further Discussion

1. Should we have a health care system based on people's needs or on their ability to pay?

2. If we created a health care system for all Americans, what, in addition to the things discussed in this chapter, would you put in an American health care system? Why?

3. What things suggested in this chapter would you drop? Why?

4. If the American people had a chance to read this chapter, what do you think their thoughts would be? Where would they agree, and where would they disagree? Why?

5. Will we have a national health care system someday? Why or why not?

6. How could we go toward a more preventive philosophy in health care?

7. Are we becoming a healthier or unhealthier society? What is your reasoning?

8. What one step do you think Americans could take to have a much healthier lifestyle?

9. How healthy is the lifestyle of college students? Why is this the case?

10. How could college students live a healthier lifestyle?

11

How Can We Solve
the Problems of Families?

Y ou might ask why I put the chapter on solving the problems of
families after all of these other chapters. I did this for a specific reason. If we can solve or ameliorate many of the social problems that we have
just discussed, we can go a long way toward solving the problems of
American families. Let me show you how solving or ameliorating these
social problems will greatly help our families.

Decreasing Inequality to Help Families

As we discussed in Chapter 3, the United States has the most inequality of
any industrialized country, and our inequality continues to widen. As you
recall, we mentioned that our chief executive officers earn the largest salaries
of any such executives in the world, whereas our workers earn some of the
lowest wages when compared with workers of other industrialized nations.[1]
Our country is the only industrialized country that does not have a health
care system for all of its citizens (Kerbo, 2006, p. 38). Other countries have
paid leaves for pregnancy that can last more than a year, whereas we have
unpaid leaves for 3 months. Other countries have more extensive child care
subsidies than we do, helping their lower income parents more than our
lower income parents. In fact, as we pointed out in Chapter 4 on poverty,
we have long waiting lists of lower income families waiting to get child care

subsidies. All of these indicators suggest that we have more inequality than any other industrialized nation (p. 27), translating into more hardship on American families. As families have fewer resources on which to survive, we should predict that these families will have more problems such as stress, abuse, and divorce.

On the other hand, as we can decrease the inequality in our country by implementing a more progressive income tax (where poorer people pay less in tax[2] and richer people pay more in tax), providing health care for everyone, and offering more child care subsidies and rent subsidies for housing, these factors together can work to decrease the stress in many low- and moderate-income families. This decrease in stress will provide the social conditions to have more stable and happier families.

Decreasing Poverty to Help Families

One of the social problems that could have a big influence on American families is getting families who are in poverty out of poverty. Probably, no one variable does more to hurt families and put more stress on them than that of poverty. If families do not know how they are going to pay next month's rent, do not have enough money to buy food for the whole month, and do not have decent clothes for their children to wear to school, all of these financial burdens can put great stress on poor and near-poor families, leading to lower self-esteem for the parents, to more arguing about finances, to more spouse abuse or child abuse, to greater strains on marriages, and to more divorces.

Although the factor of poverty is not the only factor to cause these unfortunate consequences, it no doubt plays a key role in decreasing the unity, stability, and happiness of families. If we, in our country, seriously work to get our poor people out of poverty (see Chapter 4 for specific suggestions), we can go a long way toward helping our poor families live a more stable, secure, and happier family life. As we discussed in Chapter 4, various European countries have reduced their poverty substantially by how they tax people and by the services they provide. If other countries have gotten most of their poor out of poverty, we should be able to do this too.[3]

Enough Jobs and Enough Decent-Paying Jobs

I could have put the following discussion in the preceding section about decreasing poverty, but I wanted to emphasize the importance of decent-paying jobs to the solving of problems of American families. As we discussed in Chapter 4 on poverty, there are two problems with the job structure in a capitalistic economy that play havoc on many American families. First, capitalism does not create enough jobs. Second, capitalism does not create enough decent-paying jobs for all people to live a decent lifestyle. When families cannot earn enough money to pay their bills, they will constantly be under stress, resulting in greater chances for spouse abuse, child abuse, child neglect, and divorce. Hence, the kind of economy we have is directly related to the problems our families have. As I mentioned previously, our kind of economy will probably not change soon, so that leaves us with what can we do within a capitalistic economy to create enough jobs for families to have a decent lifestyle.

As we discussed previously, creating a progressive income tax system that takes pressure off lower income families by allowing them to keep more of their take-home pay and developing a social service system that meets the unmet needs of many people will greatly alleviate the stress on many families. However, a third crucial factor in decreasing stress on American families is to create enough decent-paying jobs for people. Simply providing a living wage could help families immensely. Moreover, the combination of less tax on lower income people, more social services (e.g., child care subsidies so that mothers can work, Section 8 housing subsidies, college student loans), and enough decent-paying jobs would act to greatly enhance the livelihood of American families, and this in turn would decrease family stress and thereby decrease the chances for divorce.

Because our capitalistic economy by nature does not always provide enough decent-paying jobs, we will need to continually look for ways to create more decent-paying jobs (review Chapter 4 on poverty for specifics), and when not enough jobs are created, we will need to have other alternatives ready (e.g., guaranteed minimum income, sufficient unemployment compensation) to address the needs of families. As you can see, the creation of more decent-paying jobs will greatly help American families.

Creating Racial/Ethnic Equality to Help Families

So long as we have racial/ethnic prejudice and discrimination (individual, group, or institutional), we will have racial/ethnic inequality. These social conditions will lead to a number of families in our society not being able to have the same opportunities as other families have. These families will not receive as good of an education, will not get the chance to be as upwardly mobile, and will not get to be in occupations with more income, power, and prestige. As a result, these families will not have as much income to buy homes and accumulate wealth, will not be able to have a number of material goods in their homes, will not be able to pay their bills as easily, will not be able to send their children to good schools, and will be less likely to take vacations or be able to afford other leisure activities such as going out to nice restaurants and attending sporting events, plays, or concerts. In other words, prejudice and discrimination will result in these families having fewer life chances.

The families that endure prejudice and discrimination will, in all probability, face much more stress and many more self-esteem problems. First, they will experience the stress of not having enough money to pay their bills. Second, the inability to pay bills could also affect the parents' self-esteem. As the sociologist Charles Horton Cooley noted with his concept of the looking glass self (Cooley, 1964), we look to others (e.g., our parents, our friends, our teachers, our coaches) to see how they see and judge us, and we tend to see and judge ourselves based on how others see and judge us. With respect to parents who experience prejudice and discrimination, it would not be surprising to find that their self-esteem is affected along with that of their children.

The stress of money problems alone can put great strain on families. In his classic work *Tally's Corner,* Liebow (1967) discovered that many African American men of the early 1960s living in Washington, D.C., could not find decent-paying steady jobs. Not having the opportunity of obtaining these kinds of jobs caused many men not to want to marry because they knew they could not support their families, or if they did marry and had every intention of having lasting, stable, and happy marriages, they found that they could not get decent-paying steady jobs to keep their marriages together. In other words, Liebow found that the social structure of prejudice and discrimination had a huge effect on African American families. For example, he noted,

> In general, the menial job lies outside the job hierarchy and promises to offer no more tomorrow than it does today. The Negro menial worker remains a menial worker so that, after one or two or three years of marriage and as many children, the man who could not support his family from the very beginning is even less able to support it as time goes on. The longer he works, the longer he is unable to live on what he makes. (p. 211)

As we discussed in Chapter 5 on racial/ethnic inequality, we have made progress in decreasing discrimination in our society by creating and enforcing civil rights laws so that any American can go to a restaurant, park, stadium, or concert; can vote, hold political office, get a bank loan to buy or build a home (called a *mortgage*), get loans or grants to go to college, play on sports teams, earn an athletic scholarship; can go to graduate school, law school, or medical school; and can become doctors, lawyers, professors, and so on. So, we have made considerable progress in decreasing discrimination, and in so doing we have increased opportunities for African American, Native American, Hispanic American, and other American minorities. To the degree that we have made this progress, especially since 1950, we have helped American families from various minority groups to increase their opportunities and increase their chances for upward mobility, and hence they have the chance to have a more stable, secure, and happier home life. Consequently, as we continue to work to decrease discrimination in our society, we should observe a positive effect on minority families where there is a decrease in stress, spouse abuse, child abuse, child neglect, and divorce. Our solving or greatly ameliorating racial/ethnic prejudice and discrimination will be a key factor in helping to decrease the problems in our families.

Creating Gender Equality to Help Families

As more females get a good-quality education so that they can have a chance at being upwardly mobile and hence make more money if they become single mothers, this will help female-headed families to have decent incomes so as to have a more secure and stable family life. With the divorce rate leaving many mothers on their own to take care of themselves and their children, it is more imperative than ever before that women have a good education or training so that they will be prepared for the life of a single parent if need be. Hence, if we want to help female heads of families, our society will need to continue to promote gender equality.

Creating Good-Quality Education for All Students to Help Families

Although many American children get to have a good-quality education in the form of small classes, certified teachers, and up-to-date facilities, there are many other children who sit in large classes, have uncertified teachers teaching subjects they know little about, and go to schools that do not have well-maintained facilities and do not have enough textbooks, microscopes, band instruments, and art equipment (Kozol, 1991). In the latter situation, these school environments do not prepare children to get ahead in our society.

Due to families' racial/ethnic backgrounds and their financial conditions, many times they live in areas that have lower quality schools because the property taxes are lower, meaning that there is less money that can be allocated to having good-quality schools. As we decrease the prejudice and discrimination in society, we can help the financial conditions of these families to where they can move to areas that have better school systems.

As we discussed in Chapter 7 on education, we can also take steps to increase the quality of education in the schools where many minority families are currently sending their children. We can increase this quality by providing more money to hire more teachers to decrease the student-to-teacher ratio, to hire certified teachers, and to provide up-to-date technology and facilities. One way to get more money for schools that will be the least hurtful to the American people is to create a more progressive income tax system so that the richest people in our country pay more in taxes. Currently, the top 1% of the richest people in our country own 57% of all the wealth of our country (Kerbo, 2006, p. 34). The top 10% of the richest people own approximately 92% of all the wealth of our country. These people could be taxed more via a progressive income tax to provide for a better education for all Americans, and yet the rich still would retain much of their wealth.

Decreasing Crime to Help Families

Crime affects our society in that tax money needs to be set aside to pay for more police, judges, courts, and prisons; prisons alone can cost taxpayers $25,000 per year per prisoner, above and beyond the cost of building prisons (Macionis, 2005, p. 215). Crime also affects how much people fear being potential victims of crime. Moreover, crime affects families in that when parents are sent away to prison, families suffer not only from the lack of emotional support but also from the lack of financial support. Consequently, higher crime rates mean that more American families are disrupted for long periods of time.

If we can work to decrease crime in our society (see Chapter 8 on specific steps we can take to decrease crime), we can keep more parents out of prison and in their homes so that they can provide for their families both emotionally and financially. As you recall from Chapter 8, we can provide more legal opportunities[4] for Americans by providing more excellent public schools and trade schools and by providing more student loans for lower income students to have the chance to further their education. More educational opportunities will result in Americans having a better chance to be upwardly mobile so that they can earn decent incomes and know that they can get ahead in life.[5] The more we create the social conditions for Americans to achieve by legal means, the less incentive they will have to achieve by illegal means. With less incentive to commit crime, more families will remain intact and hence be more stable and secure.

Changing the Way We Deal With Drugs to Help Families

Currently in the United States, using and/or selling drugs such as marijuana, cocaine, heroin, and methamphetamines is a crime. If users and sellers of drugs are caught and convicted, they are given prison sentences. Many times, those found guilty are parents who are sent off to prison for a number of years. This creates havoc for these families. In Chapter 9 on drugs, I suggested that we could take an intermediate step of legalizing the use and sale of marijuana (recall that marijuana was legal for the first 161 years of our nation's history—until 1937—and has been illegal since then). This action would cut back on the parents who currently go to jail for using or selling marijuana. The benefit for these families would be that the parents would still be at home raising their children and making money to make ends meet for their families. Instead of sending these parents off to jail and putting great financial stress on these families to try to pay their bills, the act of legalizing the use and sale of marijuana would result in more parents remaining with their families and hence creating a more stable financial environment for these families.

As for drugs such as cocaine, heroin, and methamphetamines, even if we do not make them legal, our society could use more counseling, rehabilitation, education, training, and job placement options instead of simply sending parents off to prison. If we moved in the direction of a balance among counseling, rehabilitation, education, training, and job placement, on the one hand, and going to prison, on the other, we could decrease the number of parents who serve long prison sentences and therefore decrease the number of families that are devastated emotionally and financially. So, if we want to help American families, it seems that we could change our drug policy somewhat as a way to keep more American families together.

Creating a National Health Care System to Help Families

Creating a national health care system would immediately help the 46 million people in our country who have no health care coverage and would therefore help millions of families receive health care and in turn help these families to become more financially stable and healthier. This one change in our country would have a huge effect on these families.

Given that many of these uninsured families are at or near the poverty line, and that the employers of the parents of these families do not, or cannot afford to, provide health care coverage, the parents need to use what little money they receive in wages to pay for family medical expenses. Simple physical checkups and minor medical or dental expenses, such as paying to fill a cavity or to sew up a cut, will throw these near-poor families into stressful financial conditions. A major medical bill, such as paying for heart surgery, will devastate these families. Consequently, to have a health care system that provides everyone with health care would be a financial burden lifted for these 46 million Americans and would therefore greatly decrease the stress on these families and help them to maintain more stability.

Another type of family, elderly couples, would also benefit from having a national health care system because these elderly people would not need to use so much of their social security and retirement incomes to pay for prescription drugs. They would in turn not have as much financial stress on them and could live more enjoyable lives. Moreover, adult children of these elderly couples would experience less stress because they would use less of their own incomes to help out their elderly parents. Having a national health care system would therefore decrease the financial stress on roughly one of every four Americans. Consequently, a good health care system that serves all Americans would be a big step in helping many American families.

Including Homosexual Couples to Help Families

During recent years, we have had a lot of discussion in our country about whether or not homosexual couples should be allowed to marry. One way we could look at the issue is how we could help families to become happier and help relationships to become more stable. Using these two goals, what promotes more stability in relationships and what promotes more happiness in families, what could we do to achieve these two goals?

We know that a number of American males and females are homosexually oriented. Although some homosexuals have been socialized to be homosexual, there is evidence to suggest that most homosexuals are homosexual because they are born that way.[6]

Historically, our society has urged, admonished, and forced homosexuals to act heterosexual. But for most homosexuals, this was like telling a male to become a female or telling a female to become a male. So, one of the first steps our society could take to achieve the goals of creating more stable relationships and happier families is to acknowledge that most Americans, whether heterosexual or homosexual, have a sexual orientation that is strong, enduring, and highly resistant to change.

Acknowledging that most homosexuals are not changeable (just as most heterosexuals are not changeable), the next step we can take is to accept homosexuals as they are and see how we can include them more in our society rather than exclude them by discriminating against them. Note that even though I talk about homosexuals as if they were one uniform group of people, both homosexuals and heterosexuals fall along a continuum of sexuality where a number of Americans are exclusively heterosexual, other Americans are mainly heterosexual with some homosexual interest, still other Americans are oriented somewhat equally in heterosexual and homosexual ways, other Americans are mainly homosexual with some heterosexual orientation, and finally still other Americans are exclusively homosexual.[7] So, in reality, there is much more sexual variation among Americans (and among humans in general) than many Americans realize (or want to admit).[8]

As we discussed in Chapter 5 on racial/ethnic inequality and in Chapter 6 on gender inequality, when we are prejudiced and discriminate against a group of people, we tend to exclude these people, with the result that they experience many negative consequences such as fewer opportunities, lower income, less upward mobility, lower self-esteem, anger, hurt, rage, and fatalism.

In our country, given that we work to accept homosexuals as they are and that we work to include them fully in American society, it follows that we will need to create social conditions that let homosexuals live up to their potential. This raises the question of how we can do this.

One of the areas that we could improve has to do with homosexual relationships. That is, how can we make these relationships more stable and these couples happier? In addition to passing and enforcing laws that give homosexual Americans equal opportunities to get jobs, be promoted, and obtain housing, our society can take other steps to help them to have enduring relationships like those of heterosexual couples. For example, we can enact laws that allow homosexual couples to be in legal relationships. We could call these relationships *marriages, civil unions,* or something else. What we call these relationships is not as important as the fact that we create the same legal conditions for homosexual Americans to have lasting relationships as we do for heterosexual Americans. As homosexual couples have legal partnerships, they can share health care insurance coverage, own homes together, sign for loans together, visit each other in hospitals, and inherit wealth from the surviving partner. With such legal rights, homosexual couples will be able to have more stable relationships.

Although these legal rights will not ensure that homosexual couples will have everlasting relationships (just as these legal rights do not ensure that heterosexual couples will have everlasting relationships), these rights will make it easier for homosexual couples to carry on stable relationships. This set of social conditions not only will lead to more stable relationships but also will lead to happier couples.

The legal ability to have more stable relationships will also allow homosexual couples to have more stable and secure families if they have children. Whether through adoption or through one of the partners contributing his sperm or her eggs so that one of the partners is biologically related to the children, homosexual couples, by being recognized legally as a couple, will be able to have a more secure and stable environment within which to raise children. These conditions should result in less stress, child abuse, child neglect, and divorce.

It seems that whenever we change our society to include Americans that were at one time excluded (e.g., African Americans, Native Americans, Hispanic Americans, women, handicapped Americans, homosexual Americans), we give these particular Americans a chance to flourish. Once given the chance, they in turn end up making all kinds of unexpected contributions to our society and to humanity. But these contributions usually start with giving a group of people a chance by including them.

Educating Students to Help Families

Most American children in Grades 1 through 12 attend our public school system. Given such attendance, the public schools can be a venue that can teach children how they can be better spouses and parents. It therefore seems possible that we could create a curriculum that could be offered in elementary, middle, and high schools and that would discuss essential information on how to be better spouses and parents. Either a national committee of experts or a committee of experts created from each of the 50 states could develop such a curriculum. Some of the curricular topics that could be created and discussed in class are as follows: how to treat one's spouse, how to make decisions with one's spouse, how spouses can manage money, how to cook and how to cook nutritious foods, how spouses can please each other sexually, how to treat children, and how to raise children.

The marriage and family experts could arrive at a commonly agreed-on curriculum that could be used in all pubic schools where the depth of knowledge presented would progress as children advanced from elementary school, to middle school, to high school. Consequently, by the time students were 16 years old or had reached the mid-high school years, they would be better prepared to become spouses and parents than students of that age are now.

These courses will not guarantee that all marriages and families will be successful. But these courses will give students a knowledge base and reference points with which to make more informed decisions, thereby moving spouses and parents in the direction of experiencing more stable, secure, and happier home environments.

This course could be a part of the same course that we talked about in Chapter 10 on health care that will teach students how to pick nutritious foods to eat, how to cook these foods, how to get healthy exercise, and how to live a healthy lifestyle overall. Such a course not only could increase the health of students but also could prepare them to be better spouses and parents.

Some students are already taught in their homes how to pick and cook healthy foods and how to get appropriate exercise. They are also taught how to be good spouses and parents because their parents model such behavior each day. However, we know that not all students grow up in such ideal social environments. This course would help to fill the gap that these students missed as they were growing up.

Educating Parents to Help Families

We could also set up adult classes for people to learn how to be spouses and parents. These classes could be offered by public school systems, public health departments, or local governments to give adults information and advice if they are thinking about becoming spouses or parents or if they are already spouses or parents and feel that they need help or advice. As an incentive for spouses and parents to take these courses, there could be some kind of tax deduction created. Also, if spouses or parents have come before the criminal justice system due to spouse abuse, child abuse, or child neglect, they could be directed to take adult spouse and parenting courses as a part of the rehabilitative sentencing handed down by judges.

Summary of What We Can Do to Help Families

As you can see, how well families do in our society depends on how well we solve other social problems such as decreasing inequality in general, decreasing poverty, decreasing racial/ethnic and gender inequality, increasing the quality of public education, and providing health care for all Americans. Solving or addressing each of these social problems will go a long way toward helping American families to become and remain stable and happy. So, the more we can solve other social problems, the more we can solve problems in our American families.

We can also take additional steps to decrease the problems of families such as educating children in how to be better spouses and parents, educating spouses and parents in how to be better partners and heads of families, and working to create laws and social policies that include homosexual partners and homosexual families so that these kinds of families have a better chance of living more stable, secure, and happier lives.

No other chapter in this book demonstrates so clearly how social problems are interrelated as this chapter does. Consequently, as we successfully address the other social problems that we have discussed, we will at the same time take a major step in addressing the problems of our families.

Questions for Discussion

1. How much will our addressing the other social problems we have discussed so far help us to solve the problems of families?

2. In addition to what was discussed in this chapter, what else can we do to help families in our country?

3. What else can we do to decrease stress in American families?

4. What else can we do to decrease spouse abuse in American families?

5. What else can we do to decrease child abuse and child neglect in American families?

6. What else can we do to increase the happiness and stability in families so that there will be a lower divorce rate?

7. What would you do to solve problems in homosexual partnerships and families given the existing discrimination against such families?

8. Will there someday be legal relationships (civil unions, marriages, or whatever they might be called) for homosexual couples? Why or why not?

9. Of all the social problems discussed in the previous chapters, which problem, if solved, would help families the most?

10. What do you predict about our solving problems in American families during the next 10 to 20 years?

12

How Can We Solve the World's Population Problem?

W e, as humans, have a big problem. We are increasing our world's population very quickly and have been doing so for the past 100 years. Our world's population is doubling at a rate of every 54 years.[1] This would not be so bad if we had just a few million people on Earth. Instead, we have more than 6 billion people, and the last 3 billion people have been born during the past 50 years!

If we continue at the current rate of doubling the world's population, we could have 12 billion people by around the middle of this century; if you happen to be a 20-year-old college student, this means that by the time you are 70 years old, we could have another 6 billion people on Earth. These growth rates raise a number of important questions. Can Earth support an additional 6 billion people? Will we develop new technologies that provide enough food, water, and shelter for these people? Will we have enough fertile land? Will we have enough fresh water? Will we have enough housing? Will there be enough jobs for people to make money to survive? Will there be constant outbreaks of various diseases? Will many people be malnourished? Will many people starve to death?

These are important questions, yet we cannot answer these questions with certainty. To say the least, these questions are disturbing and even scary. We need to begin arriving at answers to these questions sooner rather than later. Again, if you happen to be a 20-year-old college student, your generation will need to have answers to these questions. Even if the rate of population growth would happen to slow down, where we add 3 to 4 billion people instead of 6 billion people during the next 50 years, the preceding questions are still relevant.

Unlike the social problems we have covered in previous chapters of this book, where we, in the United States, could take specific steps to remedy the problems we have in our own country (e.g., our own poverty or racial/ethnic and gender inequalities), the solution to the world's population will require the help of many countries.

With these initial thoughts in mind, let us first discuss causes of the population problem that will give us ideas as to how to solve this problem. There are two main causes on which I want to focus: the growing population of the world and the unequal distribution of resources both within and among countries of the world.

Causes of the World's Population Problem

The Growing Population of the World

Birth Rate and Death Rate

With respect to the first cause of the world's population problem, we need to consider two variables: the birth rate, which is the number of births per 1,000 people per year, and the death rate, which is the number of deaths per 1,000 people per year. If the two rates are roughly the same, as many people are dying as are being born and hence the population is not increasing. However, during the past 100 years, the death rate decreased quite a bit but the birth rate decreased more slowly. The result has been that each year a lot more people were being born than were dying, leading to an increase in the world's population.

The death rate decreased for various reasons. We started enclosing our sewage systems within pipes so that the sewage could not spread disease to people. We started growing more grain (which could be stored and used when we needed it), and this meant that people had better and more consistent diets and therefore did not die so early in their lives. We invented more medicines and vaccines, and this meant that people were not dying of diseases so often.

The birth rate also decreased, more so in some countries than in others, but not as fast overall as the death rate, hence the rise in the world's population. The reason why the birth rate decreased, but not as much or as quickly as the death rate, was that as countries industrialized, the people in those countries began to move to cities to work in factories. As families moved to cities, children were not as able to be economic contributors to the family as they were in the rural setting on farms. It therefore became more expensive to raise children. Over time, as having children became more of an economic liability, families began to have fewer children. This process happened more in industrialized and urbanized countries. Consequently, in industrialized countries, the birth rate went down faster than it went down in countries that were still mainly rural. Rural people still relied heavily on their children to help with the farming such as planting and harvesting crops and taking care of animals.

The overall result of this process of a large decline in the death rate and a moderate decline in the birth rate was that the world's population has increased greatly. This is the problem that we face in today's world.

Unequal Distribution of Resources Both Within and Among Countries

A second major cause of our population problem is the unequal distribution of resources within and among countries, resulting in people being unable to have enough resources to sustain themselves. I first discuss the unequal distribution within countries and then discuss the unequal distribution among countries.

Unequal Distribution Within Countries

With regard to unequal distribution of resources within a particular country, two factors stand out: the unequal distribution of resources because of who owns the land and the unequal distribution of resources due to discrimination by the political leaders.

Unequal Land Distribution. With respect to unequal land distribution, the problem is that in a number of countries a few people own most of the land and therefore have control over who gets to use the land and how it is used. For example, the Gini index for land inequality, which measures how much land ownership is concentrated within a few people in a country, shows that countries in Latin America such as Peru, Argentina, Brazil, Ecuador, and El Salvador have much higher concentrations of land distribution in a few hands than do other countries in the world (Kerbo, 2006, pp. 33–34). In fact, Kerbo (2006) noted, "In some Latin American countries, 10 percent of the people own about 90 percent of all the land" (pp. 33–34).

The result of this situation is that many people in these countries do not have land of their own to grow food such as rice, beans, and potatoes for their families, making it difficult for them to survive. To make matters worse, the landowners will use their land to grow cash crops such as coffee and sugar, which will be traded to developed nations. Although the landowners make a lot of profit, their fellow citizens do not have access to land as a means to survive. As these landless people have more children, many times due to the lack of having access to effective birth control, they are less able to provide enough food for their families. Hence, as landless families have more children, they have less ability to provide sufficient food for their families. The consequences for these families will be higher rates of malnutrition, disease, and starvation.

Political Discrimination and Unequal Distribution of Resources. Another factor that aggravates the growing number of people in a number of countries is that one racial/ethnic group controls the government of that country and discriminates against other racial/ethnic groups by not providing them with land, water, food, and other resources with which to survive.[2] In these situations, as these groups grow in numbers, they cannot increase their resources accordingly. The result usually is some combination of malnutrition, disease, and starvation.

Consequently, when a growing population of a country is combined with unequal land distribution or political discrimination of resources, the increase in the population becomes a problem, with probable consequences (e.g., malnutrition, disease, starvation) to follow. If, on the other hand, there is a population increase and there is a corresponding redistribution of resources to meet the increasing demand of having more people, we will not end up with a population problem—at least not at that time. However, because countries have a finite amount of resources, at least in the short term until they can discover or produce more resources, an increase in the population will mean fewer resources per person in those countries, even if they are distributed equally (and that is very rare).

Unequal Distribution Among Countries

Another problem related to the world's population problem is the inequality of resources among countries. Developed countries such as the United States, Canada, Western European countries, Japan, and Australia have more resources and can therefore provide for their people easier than can poor countries in Africa such as Sierra Leone, which has a per capita income of $460 per year, and Malawi, which has a per capita income of $560 per year (in contrast to the United States, which has a per capita income of $34,280 per year, and Germany, which has a per capita income of $25,240 per year) (Population Reference Bureau, 2003).

The poorer countries of the world frequently do not have enough resources to allow the people to live decently. This situation of having the bare minimum is further aggravated when the birth rates in these poorer countries are much higher than the death rates, resulting in quick increases in the population but not corresponding increases in access to resources such as fertile land, fresh water, food, shelter, and jobs. If the resources of the world were redistributed more equitably from richer countries to poorer countries, the increase in the population of these poorer countries would not be a problem or would not be nearly as much of a problem.

As you are probably beginning to see, the variables of birth rate, death rate, and access to resources are three important variables that influence whether or not we have a population problem. If some combination of these three variables gets out of whack, we have a problem. For example, if birth rates are much higher than death rates but the amount of resources remains the same, we will have too many people for the existing resources. If death rates plummet due to treatments to stop disease and people are having better diets but birth rates remain the same, there will be many more people being born and possibly not enough resources to take care of these people. If one group in a country has most of the resources and discriminates against other groups, the other groups will not have enough resources to live or to live decently. Moreover, if these other groups grow in population, they will have an even more difficult time in surviving. If some countries have many resources and other countries have few resources, the people in countries with fewer resources will have a harder time in surviving. As you can see, the mixture of birth rate, death rate, and unequal distribution of resources can be the recipe for disaster for many people on Earth.

What Can We Do?

There are a number of things that we, as citizens of this world, can do to solve or ameliorate the problem of rising population. We will find, however, that with each solution we suggest, there will be people who will be against that particular solution. So, the problem with solving the population problem of the world will be due not to a lack of answers but rather to the vested interests, values, and beliefs of certain groups or countries that will act as barriers to solving this problem.

Decrease the Birth Rate

A key answer to solving the population problem of our world is to decrease the birth rate of many countries. In many countries that are also struggling with poverty, mothers are bearing five, six, or seven children rather than having one, two, or three children.[3] With a birth rate that high, so many people are being born that the country cannot provide enough food, shelter, schooling, and other basic services such as clean water and efficient sewage systems. When these children are looking for jobs 15 to 20 years later, there will not be enough jobs available. So, a major solution to the population problem is to decrease the birth rate to one, two, or three children. It is estimated that "half of the married women in developing countries do not want more children but do not have access to effective methods of birth control" (Eitzen & Zinn, 2000, p. 58). Part of the answer, then, is to provide women with birth control so that they can have only the number of children they want to have.

Even if countries allow more access to birth control, there are still the challenges of overcoming vested interests and differences in values and beliefs that come into play. First, let us look at the problem of vested interests. Parents in many poor countries rely on their children in two ways to survive. First, these parents live in rural settings and need extra hands to help with the planting and harvesting of crops. Hence, they see more children as being economic assets in helping with the crops and tending to the care of various animals such as sheep, goats, cattle, and chickens. Second, when these parents are too old to farm the land, the children serve as social security for the parents by farming the land, tending to the animals, and taking care of the parents. For these two reasons—help with the farming and being a means of social security—it is in the parents' vested interests to have more children.

Let us look at the problem of values and beliefs. In a number of cultures, having children shows the rest of the community that the man of the household is a "real man." Because he is capable of having children, he has prestige. Likewise, if a woman shows that she can bear a number of children, she is seen as a "real woman" by having the ability of to carry out her role of bearing children.

As you can see, vested interests and values and beliefs can combine to cause parents to have five, six, or seven children. Also, due to diseases that children can contract, parents believe they need to have more children in case some of their children die during infancy or childhood.

This leads to the following question: How can we decrease the birth rate in countries where these conditions exist? This is a challenging question. At the individual family level, it is functional for families to have more children. However, at the societal level, it is dysfunctional for the country to have more mouths to feed and more resources to use up when the country is already strapped for food and resources. Moreover, at both the national and worldwide levels, more children mean more depletion of resources and more pollution. These depletion and pollution problems are even more pronounced in developed countries where each additional child will use a lot more resources (e.g., using more gasoline for a car and using more electricity for showers, lights, and heating, resulting in more depletion of coal, oil, and natural gas) and pollute a lot more than will a child in a developing country. So, whereas parents in poorer rural countries have vested interests in economic survival by having more children, children in more developed countries have vested interests in living a certain lifestyle that depletes resources and pollutes the environment.

We, as a world community, are coming to realize that we need to decrease the birth rate, but there are competing interests of economic survival, a desire to carry out the values and beliefs of one's country, and a desire to maintain a certain standard of living. What do we do? Given these barriers, let us discuss some actions we can take to decrease the birth rate.

Increase Funding for Birth Control

We, in developed countries, can increase the amount of money we provide for birth control throughout the world. We already know that many married women in the world do not want more children. Consequently, there is a strong desire for these women to use birth control if they simply had access to this control. Governments and countries that contribute money for birth control but do not want to use their contributions to fund abortions can stipulate that their funds go only for birth control. This money would then go for purchasing birth control devices, for training people to teach others how to use birth control devices, and for providing transportation to get birth control devices to people who live in mountains, deserts, or jungles. In other words, the goal would be to provide access to birth control to every woman in the world who wants to use it. This allows individual families to decide how many children they want, giving them freedom in their family planning.[4]

It has been estimated that it would take $8 billion per year to provide birth control for all of the women in the world who want it (Eitzen & Zinn, 2000, p. 58). Some extremely rich individuals have made more money than that in 1 year. If some individuals make this much money in 1 year, it seems well within the realm of possibility that the citizens and countries of the world could raise this much money for one of the world's most pressing social problems. The past and present Bush administrations have withheld funding for birth control programs, whereas President Bill Clinton supported funding for birth control programs.[5] So, even in our own country, there are political vested interests that work against providing enough money for the women of the world to have access to birth control.

Along with providing access to birth control, people could be taught that a by-product of providing more birth control would be that HIV/AIDS transmission rates could be decreased if certain kinds of birth control (e.g., the condom) were used but not if other methods (e.g., the rhythm method, the pill, the intrauterine device, the diaphragm) were used (Rosenfield, 2000). The distribution of condoms could be of great help for the people in many southern African countries that are currently experiencing horrendous rates of AIDS. For example, 20% of the adults in South Africa, 31% of the adults in Lesotho, 33% of the adults in Swaziland, and 39% of the adults in Botswana have AIDS (Population Reference Bureau, 2003).

Provide More Education for Women

Another way to decrease the population is to educate more women. As more women go to school and stay in school longer, they tend to have fewer children. The reasons for this are twofold. First, as women go to school longer, they begin to be more interested in careers, and this usually means that they have fewer children. Second, as women go to school longer, they tend to marry later and hence have shorter periods of time when they are capable of bearing children.

Provide Social Security for the Elderly of Poorer Countries

As a possible way to address the problem of children serving as a method of social security for parents in many developing countries, developed countries could provide foreign aid that would act as a partial replacement for children acting as a means of social security. This policy could be both expensive and controversial for developed countries, but providing such aid could be a key incentive for developing countries to decrease their birth rates. If such a program were implemented, it could take the pressure off of cities receiving so many people from the countryside, and this in turn would decrease the potential for political instability, disruption in trade, the takeover of foreign corporations, and the use of foreign troops to quell uprisings. In this way, it would be in the vested interests of developed countries to provide such a social security program in return for uninterrupted trade, protection of foreign investments, and not needing to send troops to those countries.

Forgive or Decrease Debt

Another partial solution that is also controversial would be for developed countries to forgive or decrease developing countries' debt in return for the developing countries agreeing to decrease their birth rates. The forgiving or decreasing of debt would give the governments of developing countries an incentive to find ways to decrease their birth rates. This action would be costly to developed nations in the short term, but over time the developed nations could create better relations with developing nations,[6] giving developing nations the chance to get out of their debt, increase their standard of living, and eventually be potential markets for developed nations (Stoltenberg, 1989). Also, getting developing nations out of debt would help to create more political stability, stabilize trade relations, and protect the developed countries' investment interests. Furthermore, as developing countries have a chance to develop economically due to debt reduction, they tend to decrease their birth rates (Epstein, 1998, p. 8).

A number of people also see the morality in these actions in that they would assert that it would be the moral thing to do to give poorer countries a chance to get out of their relentless poverty. That is, forgiving or decreasing the debt of developing countries not only would be a way to decrease the birth rate and eventually create new markets but also would be "the right thing to do."

Redistribute Land

Another policy that could be carried out that might not decrease population growth directly but could allow people to survive would be to redistribute land in developing countries, where many people do not have their own land, so that they can grow their own food to sustain themselves. Such a social policy of redistributing land would be an extremely controversial policy among the wealthy few who own most of the land. From their perspective, it would be in their vested interests not to give up the land so that they can continue to make profit off the land by growing cash crops. However, if there is an increasing tendency for political unrest, coups, and revolutions due to the great amount of inequality in their countries, it could be in the long-term vested interests of the landowners to give up some of their land in return for more political and social stability in their countries.

Where Are We?

As you can see, we have a problem that will not be easy to solve. Recall our theory of conflict and social change from Chapter 1 and consider how we can apply that theory to this social problem, especially with regard to values and beliefs, vested interests, and inequality. For example, part of the problem of trying to solve the world's population problem has to do with cultural, religious, and political values and beliefs that hinder or even prohibit the use of effective means of birth control that could in turn decrease the birth rate as a way of not allowing the number of people in a country to get way ahead of the country's existing resources. Another variable in our theory is how vested interests play a key role in who receives resources and who does not receive resources. A third variable of inequality, the inequality of resources within a country and the inequality of resources between rich and poor countries, plays a crucial role in the amount of resources people are able to have as the population rises.

Even though this is going to be a very difficult social problem to solve, we, as citizens of the world, need to face this problem and discuss what we can do to solve it. We have dealt with other problems and have overcome some of them, including decreasing certain diseases and decreasing the severity of certain social problems such as racial/ethnic and gender inequalities. We have also made progress in establishing world organizations, such as the United Nations, as a means to address world social problems. As you will recall from our theory of conflict and social change, two key variables on the way to solving a social problem are our consciousness that we have a problem and the ability to communicate with others about this problem. We are now doing this at this point in history. As we raise consciousness among more people and have more discussions, it seems there is a greater chance we can solve this problem.

Questions for Discussion

1. What could developing countries do to decrease their high birth rates?

2. What should developing countries do to decrease their high birth rates?

3. What could developed countries do to help developing countries to decrease their high birth rates?

4. What should developed countries do to help developing countries to decrease their high birth rates?

5. What do you predict will happen during the next 5, 10, and 20 years with respect to our growing population problem? What is your reasoning?

6. As our world's population continues to grow, what will be the consequences for developing countries?

7. As our world's population continues to grow, what will be the consequences for developed countries?

8. What consequences will there be for your own life as the population continues to grow during the next 10, 20, and 30 years?

9. What do you think are the possibilities for land redistribution in Latin American countries?

10. What do you think are the possibilities for rich countries forgiving or decreasing the debt of poor countries?

13

How Can We Solve the World's Environmental Problem?

We have a big environmental problem. We have been polluting our planet's air, water, and land; depleting its resources; and accumulating a lot of waste for which we need to find places to store. Why have we been doing these things? There are four major reasons. One major reason relates to Chapter 12 in that we have so many people on Earth today (more than 6 billion) and all of these people need food, water, clothing, and shelter. In addition to these minimal needs to survive, millions of people in developed nations have cars, large homes, air-conditioning, heated houses, washers, dryers, heated water for bathing, refrigerators, stoves, and so on. Hence, having many people on Earth and having many people with a high standard of living means that we will pollute a lot, deplete a lot, and build up huge amounts of waste that need to be stored.

A second major reason why we have a big problem of polluting, depleting, and storing of waste is that during the past 200 years, we have gone from an agricultural way of life to an industrial way of life. Instead of "living off the land," growing grain to eat, and tending to farm animals, we built factories and machines and created a new status called the *factory worker* to produce all kinds of products to consume, as a stroll through a typical shopping mall will show. To create these products, we have used a lot of resources and polluted the air and rivers with many pollutants.

The key to this industrializing, and hence polluting and depleting, has been the creation and development of modern capitalism. Simply stated, capitalism produces products to make profit. Thus, the more products people produce and sell, the more profit owners of factories and businesses will make. It is therefore in the vested interests of those who own factories to produce as much as they can and to sell it so as to make as much profit as they can. This process has had, and still has, the end result of using up a lot of resources and polluting a lot.

As of this point in human history, although capitalism has helped to create the highest material standard of living that the world has ever seen, it has at the same time unintentionally increased the rate at which we pollute, deplete, and have waste storage problems (recall our discussion of latent dysfunction in Chapter 1). As former Vice President Al Gore concluded in his book *Earth in the Balance*, "Human civilization is now the dominant cause of change in the global environment."[1]

A third major factor that has caused our environmental problems is that we have created an ideology within capitalistic developed nations, and that is spreading more and more to developing nations due to television, computers, e-mail, and the Internet, that people want an ever higher material standard of living (Eitzen & Zinn, 2000, pp. 91–94; Ritzer, 2005). A higher material standard of living means producing more material goods—cars, houses, washers, dryers, refrigerators, stoves, water heaters, heating and air-conditioning systems for homes and offices, and so on. All of this means that we need to use more resources to produce more products, resulting in the further depletion of many resources. As we produce and consume these products, we pollute the air, water, and land more and have more waste left over.

People in developed nations have become accustomed to ever higher standards of living. For example, in the United States, we make up 4.5% of the world's population, but we consume 25% of all the oil, coal, and natural gas (Eitzen & Leedham, 2001, p. 208) that is consumed in the world so that we can enjoy driving cars and heating and cooling our homes and places of work. So, our country alone uses up a lot of the world's resources to have and maintain a high material standard of living. Our country also does a disproportionate amount of polluting to have a high material lifestyle. Again, we are only 4.5% of the world's population, yet we add 20% of the carbon dioxide produced by humans to the world's air, producing what has become known as the "greenhouse effect," which scientific evidence indicates is warming up Earth and starting to produce climate change. As you can see, we have a high standard of living, but we deplete and pollute huge amounts to enjoy our high standard of living. This raises an important question: Should we, as Americans, be more responsible for solving the problems of depletion and pollution given that we are responsible for so much of both?

The fourth major cause of our environmental problem is that, increasingly, people in developing nations are noticing how well people in developed nations are living and want to have some or many of the same amenities as do people in developed nations. They too want washers and dryers, air-conditioning, televisions, cars, computers, cell phones, and so on. The result has been increased depletion, increased pollution, and increased accumulation of waste in these countries as well.[2] We are currently creating a new worldwide ideology where people, in increasing numbers, want and expect a higher material lifestyle. So, it appears that the increasing pressure to produce more goods and services to have a higher material standard of living for more than 6 billion people will mean that we, as humans, will continue to face environmental problems now and in the near future.

Consequences

One of the most dangerous consequences of our polluting the environment is that of putting huge amounts of carbon dioxide in the atmosphere as the result of burning fossil fuels such as the burning of oil and gas by industries and automobiles. The result is the warming of our planet and the possible change of Earth's climate that could wreak havoc for the growing seasons, the amount of rainfall needed for crops, the rising level of the oceans[3] with the endangerment of many coastal cities throughout the world, and other problems.[4] Data, for example, indicate that "the 12 warmest years in the 140-year record have all occurred since 1983" (Gore, 2000, pp. xiii–xiv). Because we, as humans, have caused this problem (recall our theory and how we socially construct the social conditions in which we live), we have got to find a way to get ourselves out of this potentially life-threatening situation given the drastic changes in the climate that could negatively affect our ability to grow the food we need to survive.

Climate change will cause certain areas of the world that had produced sufficient food to produce less food, causing malnutrition and starvation. For example, scientific evidence now suggests that the spewing of sulfur dioxide in the air by industries from the United States, Canada, Europe, and Asia caused the decrease in rainfall in Africa, creating malnutrition, starvation, and famine (Verrengia, 2002, p. A2). This condition will create the incentive for people to move to find places to grow enough food. Hence, as we see more global warming, we probably not only will see more starvation and malnutrition in certain areas of the world but also will see more people from these areas migrating in search of a way to feed themselves (Gore,

2000, p. 73; Intergovernmental Panel on Climate Change, 2001). Moreover, with the rise of global temperatures due to more carbon dioxide in the air, the ocean levels will rise. One third of the world's population lives near coastlines (Gore, 2000, p. 74; see also Hunter, 2001). The cities these people live in will be in danger of being flooded. The result will be that many people will need to migrate inland from the coastlines. Consequently, such migration will cause the stretching of resources wherever these people go. They will need new housing, land, and access to water. To tie this migration process to the rapid rise in the world's population is to make us ponder how we, as global citizens, can and will respond.

Also, another tie to Chapter 12 is that there are two major sources of producing carbon dioxide: the burning of fossil fuels and the exhaling of carbon dioxide by more than 6 billion people (Gore, 2000, p. 93). Instead of having a few million people exhaling carbon dioxide (as was the case more than 100,000 years ago) or even having 1 billion people exhaling carbon dioxide (as was the case 100 years ago), we have many more people exhaling carbon dioxide today. So, the mere increase in Earth's population is another factor that could cause a major change in Earth's climate—something we do not want to occur.

We have also polluted a lot of the fresh water on Earth, meaning that more than 1.7 billion people do not have safe drinking water (Gore, 2000, p. 110; see also Pimentel et al., 1998). That is nearly one third of all the people on Earth! Moreover, one half of the world's population is in danger of having contaminated water because human waste is not properly treated before it is dumped into rivers. One consequence will be the rise in diseases such as cholera and typhoid (Gore, 2000, pp. 109–110). Industry has also caused a lot of water pollution because many industries have deposited their wastes in rivers. In a number of instances now and in the past, factory owners wanting to make profit did not take responsibility for the waste they produced while making their products. More profit was made, but at the expense of the environment in general and at the expense of having fresh water in particular. As more developing countries industrialize, they too will probably have more problems of having enough fresh water because their industries might not want to take responsibility for the pollution they produce.

Also, besides the cutting down of large portions of rain forests due to the rising populations of developing countries and their need for more agricultural land to grow crops to sustain themselves, the option that governments of debt-ridden developing countries have to pay off their debts or just to pay the interest on their loans has caused them to cut down the trees from their rain forests to sell as lumber. The consequence of this process has been the continual destroying of the rain forests. Yet the rain forests help to produce fresh water and produce oxygen and also consume some of the excess carbon dioxide that humans produce in burning fossil fuels and exhaling.

Another potentially dangerous consequence of our cutting down the rain forests of the world is the loss of thousands—even millions—of species of plants, animals, and insects. As yet, we do not know all of the consequences for the environment of the loss of these species. Hence, Ehrlich, Daily, Daily, Myers, and Salzman (1997) made a good point when they stated, "Until science can say which species are essential in the long term, we exterminate any at our peril" (p. 101).

In the capitalistic industrial world that we currently live in, and with more than 6 billion people wanting a higher material standard of living and being socialized by advertisements to want a high standard of living, we have come to the point where we have created, and are increasingly creating, a lot of waste and do not know what to do with it all (Gore, 2000, p. 151). Landfills are filling up. States are trying to send their waste to other states. A huge consequence of our modern industrial world is the buildup of waste—and we do not know what to do with it. Along with this consequence of a huge buildup is that when it is stored, it is usually stored on the cheapest land, typically where poorer people and minorities live nearby (p. 149). Hence, poorer people and minorities are the ones who have needed to live nearest the waste sites.

What Can We Do?

I want to say some general things about how we need to view the environmental problem before I get into specific actions we can take. In considering the environmental problem, we need to think in long-term ways rather than short-term ways (Marchetti, 1986, p. 1). If we think only in the short term, we are more likely to think of our own vested interests of our profit, our convenience, and our standard of living. If we think that way only, we will not think about what is for the good of our world now and in the future.[5] By thinking short term, we will be less likely to think about what is good for the community (local, national, or global) and more likely to think about what is good for our vested interests and ourselves. By thinking more long term, we will more likely think about what is for the good of our community—be it local, national, or global. Consequently, we will be much more likely to think about our environmental problems and how they currently affect us and how they will affect us more and more if we do not address them. By also thinking more long term, we will consider coming generations besides our own generation and their time on Earth (Gore, 2000, pp. 191, 195, 269, 333; Jan, 1995). Our society, and our world, have largely emphasized short-term gains, especially in a capitalistic economy where we emphasize profit and keeping costs to a minimum (and hence not wanting to include the costs of pollution and waste storage) rather than considering long-term costs such as climate change and ozone depletion. More and more people are coming to the realization that we must consider the long term, what is good for the global community, and what is good for future generations if we want to survive as a species. This will require us to think, plan, sacrifice, and change our ways of living. As our theory of conflict and social change points out, we, as humans, socially construct this social world, and it is up to us to socially reconstruct what we do if we want to survive by solving this environmental problem.

One step we will need to take is to become more educated about this problem (Gore, 2000, p. 223). Many people in our country, and throughout the world, do not realize the seriousness of our environmental problem. Hence, part of the solution is to make them more aware of the problem and to get them more educated about the problem (recall our theory of conflict and social change, the accompanying causal model, and the important part that consciousness or awareness plays).

Along with those who do not know about the problem, and therefore need to be educated, are those who do know but want to deny the existence of the problem or the seriousness of its effects (Gore, 2000, p. 223). Many times, they will have vested interests of profit, short-term gain, and/or their jobs at stake. These are the people who will many times put up roadblocks when the rest of us create more education about the environmental problem and create social policy to address the problem. We must find ways to give these people incentives to want to address the environmental problem or at least accept the notion that we must address this problem whether we want to or not. It will not go away by denying its existence or by denying that it is becoming a bigger and bigger problem.

The oceans and the air are common property that we all share. If air and water pollution always stayed in the country that produced it, the nation producing this pollution could take the initiative and responsibility to clean up the pollution it produced. Because air and water pollution does not stop at state boundaries but rather continues throughout the world, we, as citizens of this world, will need to work together to clean up our oceans and air. As Murshed (1993) stated, "We could view the environment—the oceans, the atmosphere—as global common property (public good), and then the problem would be to prevent excessive use or misuse of the global common, which would once again require international cooperation" (p. 43).

Probably, we will eventually need to agree on some worldwide goals for the planet. We have already been moving in this direction by attempting to decrease carbon dioxide and chlorofluorocarbon emissions. We will probably need to make saving our environment one of the main goals of our world community. Within this goal, we will need to agree on more specific goals such as decreasing carbon dioxide, sulfur dioxide, and chlorofluorocarbons; increasing the amount of rain forests; and finding new, better, and safer ways of storing hazardous waste.

In addition to agreeing on goals, we will need to negotiate agreements among nations as to what we want to do and how we want to accomplish our goals. This will not be easy. As you might predict, national vested interests will collide with international environmental policy. For example, President George W. Bush backed out of the Kyoto Protocol, an agreement among nations to decrease carbon dioxide emissions ("Out of Denial," 2002, p. A6). Because the United States produces 20% of all the carbon dioxide and is the wealthiest and most powerful nation in the world, when it backs out of an agreement, the original agreement can be hurt considerably. So, for international agreements to work, rich and powerful nations will need to be a part of these agreements.[6]

Creating common environmental goals and reaching international agreements will be helped along greatly as more leaders of various countries take steps to take our common environmental problems seriously. This will not be easy for many leaders of nations because to remain popular and get reelected, these leaders will focus more on their nations' immediate problems such as how their nations' economies are doing. Problems of internal political unrest will greatly distract leaders from focusing on the environment. Also, wars and conflicts with other countries will deter leaders from focusing on the environment. It will not be easy to get leaders of nations to focus on something that seems to be distant and therefore does not seem so urgent. But the mounting evidence of environmental degradation will increasingly present national leaders with the harsh reality that something must be done.

At this point, serious goal making and agreement making will take place.[7] One factor that may speed up this process a bit is if some national leaders, especially from the more powerful countries, promote the environment as part of the overall national and international policy.

In thinking about what we can do, a key factor to keep in mind in all of the social problems that we face is that we will need to (a) plan and think ahead, (b) consider the consequences, (c) experiment with new social policy, and (d) carry out social policy that is effective.[8] If we, as humans, do not do this, we invite chaos by not taking charge and seeing what we can consciously do with each social problem. For example, recall the utter chaos that occurred after Hurricane Katrina hit New Orleans, Louisiana, and the surrounding area in August 2005.[9]

Each day, each month, and each year, the problem with the environment gets worse. Because the environment is a problem that is less visible than other social problems, we are less likely to realize its effect (e.g., more carbon dioxide in the air causing more global warming, more chlorofluorocarbons in the air causing a larger hole in the ozone layer and allowing more cancer-producing ultraviolet rays to bombard Earth).

Also, as the environmental problem grows, there could be a point where the environment has gotten so bad that we are not able to take enough action to reverse the problem in time for humans to survive. Consequently, the sooner we take action on this problem, the more likely we can control it and solve it. The one thing that we cannot afford is to put off facing the environmental problem because this problem and its consequences can only get worse, with the worst scenario being that we can no longer control it.

In thinking about our environment and what we, as humans, need to do to survive on Earth and enjoy our time here, we need to address and solve three problems: pollution, depletion, and storing the wastes we cannot recycle. By "we," I mean two groups of people. First, we, as citizens of the world, need to address these problems. Second, we, as citizens of the United States, need to address these problems. The reason why I distinguish between these two groups is that we in the United States, along with people in other developed countries, have a special obligation to address these problems because we contribute to them disproportionately (recall the earlier cited statistics that Americans make up only 4.5% percent of the world's population but consume 25% of the fossil fuels of the world, such as oil and coal, and add 20% of all the carbon dioxide produced by humans).

The implication of these statistics is that we, in developed nations, especially need to contribute to solving these environmental problems because we contribute to them disproportionately. Also, those of us in developed nations, given our greater wealth, have a greater capacity to do something about the environment.

As an overall partial solution to our environmental problems, we need to stabilize the population of the world (Boeker & Van Grondelle, 2000, p. 82). Given the current set of social conditions, as we have more people on Earth, we will have more pollution, depletion, and buildup of wastes. This is a good example of where one social problem (increasing population) contributes to another social problem (increasing environmental problem). Consequently, if we can solve one problem (increasing population), we can help address another problem (increasing environmental problem).

Depletion: What We Can Do

As we create a higher material standard of living, we use up oil, coal, wood, metals, land, and other natural resources. With more people on Earth and more industrialization to produce more goods to have an ever higher standard of living, we use more and more natural resources to turn these resources into the goods we want. One of the problems with this is that we could run out of a number of these resources.

One answer to this problem is to recycle the resources we have already used and to use them over and over. The trick is to know how we can do this and then make it profitable, if possible, for companies to want to be part of the recycling process. As a way to make this happen, the government can do more research to find new ways to recycle and reuse our resources. Also, we can give tax incentives to new businesses that want to go into the recycling business as a way to make money.

An additional way to solve the problem of depletion is to use different resources in place of the resources we are currently using. For example, instead of relying on oil and coal to produce much of our energy, we could do more research to find more and better ways to use solar, water, and air power that would not put carbon dioxide and sulfur dioxide in the air. These kinds of energies are not used up, can be used over and over, and do not hurt the environment. If we especially want to address the problems of global warming and acid rain, we will need to use less oil and coal and to use more solar, water, and wind power.

How can we do this? We can take some of our government tax money and put it into creating more ways to use solar, water, and wind power. For example, Boeker and Van Grondelle (2000) stated, "The present period should be used to introduce renewables on a large scale, both by stimulating research and development and by adapting the energy infrastructure" (p. 80). We can also have local, state, and federal governments create tax incentives for business, industry, and homeowners to use these kinds of energy sources.

A big problem that will stand in the way of using more solar, water, and wind power is the potential loss in profits by oil and coal companies and the potential loss of jobs these industries provide people. As a remedy for this, and as a way of diffusing opposition by the oil and gas industries, we could give these industries tax incentives to create a profit-making industry based on the selling of solar, wind, and water power. In other words, these oil and coal industries could begin to switch over to new kinds of power to market. If there were employees who lost their jobs as we cut back on oil and coal production, the government could help people to find new jobs and also train them.

Because the use of oil and coal causes great problems for our environment, we need to become more serious about using nonpolluting and nondepleting sources of energy. If the climate and overall environment were not affected so much, we could continue to go on as we have been doing. But that is not the case. We are fooling ourselves if we think that there is not much wrong with the environment and that we can go on as we have been doing.

A third partial solution to the environmental problem is to redistribute land, especially with regard to the depletion of rain forests and the use of much of a country's land to produce cash crops for export instead of producing crops that feed the people of that country. If more poor people of developing countries had enough land to farm to grow enough food to sustain themselves, the pressure to cut down the rain forests would be less.

If land redistribution could occur, there would also be less malnutrition, less starvation, and less disease because people could grow their own food, have better diets, and (as a result) be healthier. As you might predict, land redistribution, as a partial solution to the problems of depletion of rain forests and lack of food, is very difficult because it would require the wealthy people of these countries to give up some or a fair amount of their land. Even though the wealthy people could still keep a fair portion of their land, they might not like this solution. As an incentive to get land redistribution to come about, the developed countries could agree to decrease or abolish the debt load on these countries. Also, by the developed countries forgiving part or all of the debt of these developing countries, these poorer countries would have less pressure on them to farm mainly cash crops for export that take much land away from the production of sustainable crops for the people of these countries.

Another important partial solution that the developed countries could sponsor for developing countries is to provide the money for developing countries to plant millions of new trees to replace the trees that have been used up for firewood. As you might already know, as there is a disappearance of trees and entire rain forests, there is less rainfall and hence less fresh water for the people to drink and use in farming. This has happened in Haiti, where many trees have been cut for firewood. As a result, Haiti has had less and less rainfall. With increasing populations in developing countries with decreasing rainfall, this is a recipe for disaster. Hence, we need to plant trees to stop and reverse the process of forest depletion. This could become one of the activities of the Peace Corps. As an excellent example of what can be done, Wangari Matthai from Kenya created a social movement to plant trees that resulted in the growth of 7 million new trees, helping to stop erosion, create more rain and fresh water, and use up the carbon dioxide in the air (Gore, 2000, p. 324).

Another partial solution is for our government to be the role model in terms of conserving our resources. For example, local, state, and national governments could use only recycled paper and longer lasting light bulbs (Gore, 2000, p. 350). They could establish higher mileage requirements for cars, sport utility vehicles (SUVs), and trucks, slowing down the depletion of oil and at the same time decreasing the carbon dioxide emissions produced in our country (p. 350; see also Renner, 2000).

Pollution: What We Can Do

We can continue to write and enforce laws dealing with pollution. Consequently, this requires that the government be involved in this process. Given the nature of capitalism and the desire to make a profit by keeping down costs, if owners and managers of business and industry are left alone, many times they will not take responsibility on their own to pollute less because doing so means increasing their costs and decreasing their profits. As a result, the government will need to continue to create and enforce antipollution laws and to provide tax incentives and tax penalties for business and industry to decrease their pollution. This will require us to have a stronger Environmental Protection Agency than currently exists, with sufficient funds to carry out any needed enforcement.

A second way to get business and industry to decrease their pollution is to provide them with money to conduct research to decrease their pollution (Gore, 2000, p. 320). Also, other research entities, such as universities and private research organizations, can be given more funding to find new ways to decrease pollution. For example, Romm (1991) pointed out that during recent years, "the cost of wind-generated electricity has dropped by 80 percent, to under seven cents per kilowatt hour, which is competitive with some of today's conventional power sources" (p. 32). Also, the cost of solar energy has decreased to such an extent that it too is becoming more competitive with the cost of using coal and oil (p. 34). As universities and private research organizations are given more funds to find new ways to decrease pollution, their new discoveries will provide ways to compete financially with current systems that pollute.

Another way to decrease pollution is to address the carbon dioxide emissions by cars, SUVs, and trucks. Higher standards can be placed on all of these vehicles. The federal government can give tax incentives to car companies to create lower levels of emissions and can levy tax penalties on those that do not make any progress. Also, the government can provide funding for research to car, SUV, and truck companies; universities; and independent research institutes to create ways to decrease carbon dioxide emissions. Realistic deadlines could be set up for car companies to retool to meet new standards. In other words, tax incentives, tax penalties, and more research together can be ways to get these companies to meet these deadlines.

Another way to decrease pollution is to use more mass transportation in the larger cities of our country. Because the cities and states in our country are not in great financial shape, the federal government will need to help in this effort. By having the latest technology applied to creating mass transportation systems throughout the United States, we, as a country, could make a large dent in the amount of carbon dioxide emissions we produce. We have a greater responsibility in our country than in other countries to decrease the carbon dioxide emissions. We can look to Japan, Germany, and France as our role models. Japan has a steel-wheel-on-steel-rail train that goes 130 miles per hour (Moberg, 1993). Japan is also using "maglev" technology, where magnets are used in place of wheels and the train can go 300 miles per hour (p. 14). Germany and France are creating trains that can go 200 miles per hour and that will link major European cities (p. 14).

We could also cut carbon dioxide emissions by using more rail transportation to transport goods around the country instead of the thousands of trucks that produce a lot of greenhouse emissions. The problem with this is that many of our factories are located next to interstate highways and have increasingly moved away from rail lines. Although Boeker and Van Grondelle (2000) suggested this as a partial solution (p. 85), it does not seem too realistic for the United States, as compared with European countries, because so many of our plants are now placed near interstates. We might need to look into this matter more and to consider the pros and cons. Given that such a move would put a lot of truck drivers out of work, and given that there could be a huge expense in building new rail infrastructure, it might be more logical to build new truck engines that produce little or no greenhouse emissions. More research on the pros and cons of truck use versus rail use needs to be conducted.

We have emphasized cars and interstates so much in our country that we have not given other modes of transportation a chance. If we could look more seriously at other modes of transportation, as well as look at what other countries are doing, we might be able to come up with a new transportation system that is much more energy efficient and environmentally friendly. The federal government could give more research money to car companies, universities, and independent research institutes to create a much less pollution-producing and resource-depleting transportation system.

One way to address the high carbon dioxide emissions of cars, SUVs, and trucks is to change from the current gasoline-using engines to other kinds of engines. For example, we are beginning to see cars that use a combination of gasoline-consuming engines and electricity from batteries to propel cars, where the gasoline engines recharge the batteries. This promising technology, along with the use of new kinds of fuel such as ethanol, could cut carbon dioxide emissions substantially.

Another trend that may continue, and that could decrease our air pollution from cars, is the trend of working at home. If businesses find that they can continue to cut costs by not having so much office space and can cut costs of salaries, retirement benefits, and health care benefits by not hiring full-time employees, the search for higher profits through lower employee costs could have the unintended consequence of decreasing carbon dioxide emissions. If this trend continues during the coming years, it could be a mixed blessing. We will have lower pollution. However, a number of people could have lower salaries, fewer retirement benefits, and fewer health care benefits. We will need to continue to watch how this trend continues and what the consequences will be.

At the international level, given increasing evidence of global warming due to increasing amounts of carbon dioxide in the atmosphere, we will need to work as a world community to reduce carbon dioxide emissions. The United States initially agreed to a voluntary agreement, but President Bush rescinded that agreement. However, it now appears that, given the mounting evidence of global warming, we, as a world community, will need to make some type of global agreement.[10]

One of the things that we, as world citizens, need to do is stop the destruction of our rain forests and begin to replenish these forests. This is easier said than done, but we will need to do it for a number of reasons. To decrease the carbon dioxide that is causing the warming of our world, we will need to have more forests to use up carbon dioxide. Also, for people to have sufficient water for drinking and water for their crops, we need more forests to produce more fresh water.

At least three things can be done to replenish the rain forests. Developed countries can provide seedlings to poor nations, such as Haiti and Ethiopia, that have lost much of their forests and hence much of their rain that produces fresh water to drink and grow crops. These countries will, in all probability, not be able to afford to purchase these seedlings. Consequently, if developed countries do not do this, it will probably not get done. A second action developed countries can take is to forgive much of the debt that poor countries have so that they can use what little money they have to dig themselves out of their poverty. With less debt or no debt, developing countries can use the money they would have paid for interest on their loans to buy the seedlings themselves or invest in their countries in other ways (Murshed,

1993, p. 41). Also, with more money available, they will be more likely to use pollution control devices. A third, more long-term solution is to provide these countries with birth control devices at low cost or no cost, the training on how to use them, and the transportation to get these devices to inaccessible areas. Hence, this should begin to slow down the population growth, and this in turn should help to put less pressure on people cutting down the rain forests over time because there would not be as much pressure to acquire more land to produce more food and to consume more timber for firewood. As you can see, the solving of one social problem will help to diminish another problem in that decreasing the population will take pressure off people cutting down the rain forests, and this in turn will help to solve the problems of too much carbon dioxide and not enough fresh water.

Storage of Waste: What We Can Do

One of the key actions we can take in the area of decreasing our waste is to connect research entities to business and industry in terms of how business and industry can find profitable ways to use the waste products they produce. In other words, one way to solve our waste problem is not to have so much waste left over. One way to decrease our waste is to change the way we produce goods so that we reduce the overall waste and then find new ways to recycle the waste we do produce (Gore, 2000, p. 146). A good example of this is how German electrical plants that used coal to generate electricity took their waste and turned it into a marketable product. The process worked as follows. As the air pollution was created from burning coal, these plants used what are known as *scrubbers* to trap sulfur dioxide and other air pollutants by spraying a mist of water and limestone over the smoke before it left the smokestacks. The resulting waste, known as *sludge,* was then processed to make wallboard for homes and other buildings (Moore, 1995). In other words, the German plant decreased the amount of waste in the production process. So, a new norm, and possibly a new law, could be created, where whenever we create a new production process, we need to include within our overall planning of this new production process what we intend to do with the waste left over so as to ultimately decrease our waste.

The principle is clear. How can we help business and industry to take their waste and not only decrease it but also make something marketable with it? This will require new research on how we can invent new ways to decrease and market waste. If we cannot find a profitable way to market it, how can we at least find ways to decrease the amount of waste and store it safely? We can give companies tax breaks for doing their own research and can give universities and other research entities funding to do research. Innovation will seldom occur in this area unless we provide incentives for business, industry, universities, and research institutes to do research on these matters. This means that we, as taxpayers, will need to provide tax revenue for more research.

There is a lot of hazardous waste in the form of different chemicals produced by business and industry, as well as a lot of radioactive waste produced by nuclear power plants, that needs safe storage. The most recent solution that has been suggested is to store hazardous waste in underground caves that are fairly inert in their physical stability. This is not a perfect solution, however, because the hazardous material must be transported to the new storage site (Carroll, 2002a). There is a concern that trucks and trains carrying this waste could have accidents (although the evidence indicates that the material so far has been transported safely).[11]

If we use this method, we will be transporting much more waste in many more trucks and trains than we have transported in the past. Hence, there will be a greater chance that accidents could occur. Probably, during the coming years, we will need to see how we can make the transportation process safer if we choose to transport waste to one site. Currently, our government is considering transporting our hazardous waste to caves 1,000 feet underground within Yucca Mountain, located 90 miles north of Las Vegas, Nevada (Carroll, 2002a, p. A10). The U.S. Senate endorsed using Yucca Mountain as the collection site (Carroll, 2002b), and President Bush signed the congressional bill into law during the summer of 2002 (Associated Press, 2002).

Another potential problem we currently face if we do go ahead and store this waste in these underground caves is that if these substances leak, they could get into the ground water and contaminate, and hence devastate, much of our water supply. So, this potential problem calls for additional research into the matter both in coming up with safer storage places and making the substances themselves in some way inert, possibly by combining them with other substances. It appears that we need to do more research on the matter.

Another partial solution to the nuclear waste problem is to create fusion reactors. Kahn and Brown (1975) asserted, "The fusion reactor is nearly free of radioactive threats . . . [and] would not leave any radioactive waste to be disposed of directly or indirectly" (p. 334).

One of the solutions that our country has been trying more during recent years is that of burning solid waste in incinerators. There are, however, both good and bad consequences of this partial solution to reducing our volume of waste. The good is that we can reduce the volume of waste by 90% (Gore, 2000, p. 157). The bad is that we create more air pollution by putting into the air a number of poisonous substances such as dioxins, arsenic, and mercury (p. 156). So, although we may be forced to burn some of the waste in the short term, this is an unsatisfactory solution in the long term because it adds poisonous waste to our air. We, as citizens of the world, will need to do more research on this matter as well.

Conclusion

To carry out the preceding measures will not be easy. As I mentioned in previous chapters, vested interests will play a strong role in preventing or slowing down the these solutions. This will happen in the United States as well as in other countries. Even though The Netherlands is seen to have some of the most progressive environmental policies in the world, it too has faced the problem of vested interests of corporations, the vested interests of making more profit and keeping down costs, and the influence of those who have power in economic matters to hold sway over those who have concern for environmental matters. Barriers will continue to be placed in front of those who seek a healthier environment. Consequently, the general public and those who have economic interests will need to be convinced that these proposed solutions will need to be carried out if we want to improve our environment and have a sustainable place for humans to live.[12]

In concluding this chapter, we might consider what a *New York Times* columnist said about what he hoped President Bush would say in his January 2006 State of the Union speech (Friedman, 2006, p. A9). He wished that the president would say that the "direction in which America needs to go is obvious: toward energy independence" (p. A9). Friedman hoped that the president would add that "we must impose the highest energy efficiency standards on our own automakers and other industries so we force them to be the most innovative" (p. A9). He said that we need to encourage a lot of young people to study math, science, and engineering so that we can "make oil obsolete" (p. A9). So long as we depend so much on oil, the United States and other countries that depend on oil will need to coddle to countries that own the oil even though these countries may have unethical dictators and policies that we do not support. Friedman therefore urged that the president send to Congress the Energy Freedom Act, where there would be great disincentives to produce SUVs and great incentives to produce energy-efficient cars as a way not only to save our environment but also to save our deeply hurting auto industry (recall from Chapter 10 on health care that General Motors and Ford announced plans in December 2005 and January 2006 to lay off 30,000 workers from General Motors and up to 30,000 workers from Ford) (Durbin, 2005; Maynard, 2006). With such incentives, this "will force Detroit to out-innovate Toyota" (Friedman, 2006, p. A9). Considering what we have said in this chapter, Friedman's call for President Bush, or any future president, to seriously address the problems of our environment seems timely.

Questions for Discussion

1. What else could we do in our country with regard to improving the environment?

2. What do you think we should do in our country with regard to improving the environment?

3. What do you predict we will do in our country with regard to improving the environment during the next 5, 10, and 20 years? What is your reasoning?

4. What could other developed countries do with regard to improving the environment?

5. What should other developed countries do with regard to improving the environment?

6. What do you predict other developed countries will do with regard to improving the environment during the next 5, 10, and 20 years? Why?

7. What could poor countries do with regard to improving the environment?

8. What should poor countries do with regard to improving the environment?

9. What do you predict poor countries will do with regard to improving the environment? Why?

10. During the next 10 to 20 years, will nations come together more to not only discuss problems of the environment but also make binding agreements?

14

Solving Our Social Problems

Predictions and Conclusions

W hat can we say about the future of solving our social problems, especially from a sociological point of view? Based on what we have discussed in previous chapters, we can make a number of predictions. As we discuss these predictions, keep in mind our theory of conflict and social change, the accompanying theoretical propositions, and the causal model as a way to understand what will happen during the coming years vis-à-vis our social problems and how and to what degree we will solve these social problems.

A Number of Similar Social Conditions

First, many of the social conditions for each of the social problems we have discussed have remained the same. For example, our economy is still capitalistic and continues to globalize. Our political system is still a two-party system with mainly conservatives and liberals deciding social policy. Our main values are still the same; that is, most Americans value equal opportunity, individual freedom, and justice.

Defining a Social Condition as a Social Problem

In terms of defining a social condition as a social problem, we should see that the defining of what is and is not a social problem will continue to be the key to the origin of a social problem (Blumer, 1971). At times, we will observe a group or an organization bringing what it sees as a social problem to the attention of the general public and political authorities. At other times, we will see a political authority, such as the president or some other highly visible government official, attempt to turn a social condition into a social problem by trying to convince the general public that such a condition is indeed a social problem (Ferrari, 1975). Either way, someone or some group will socially construct what social condition that person or group perceives should be seen as a social problem.

Legitimation

As a certain social condition gets defined as a social problem, we should observe the process of legitimation occurring, where people communicate with others to try to convince them that a certain social condition is in fact a social problem. People will increasingly question the legitimacy of the existing set of social conditions, with the possibility that political authorities will begin to pay more attention to the problem, deliberate on it, and seriously consider it as a social problem. It is at this point in the process of a social problem that we will notice a filtering process going on, where these officials will select a few of the many social conditions capable of being defined as a social problem and dub them a "social problem" for reasons we discussed in Chapter 1 such as which social problem is seen as most urgent and serious, which problem affects the vested interests of the rich and powerful, and which problem gets the most media coverage. We should also observe intense competition by various groups and organizations to get "their" social problems accepted as legitimate social problems in the eyes of the general public and political officials. With limited resources, we should also predict that it will not be possible for all social problems to be addressed at once and with equal attention and allocation of resources.

Values and Vested Interests

Values and vested interests will play major roles in the life of a social problem. We, as sociologists, will need to ask the following questions for each new social problem. What values are being promoted in connection with this social problem, what values are not considered, and why? Whose vested interests are at stake in terms of who is going to gain or lose money, power, and prestige? Our being aware of the values and vested interests involved in a certain social problem will help us to understand the nature and process of the social problem. We will constantly need to look behind appearances, as Berger (1963) stated in his book *Invitation to Sociology,* to discover the values and vested interests that are being promoted.

We, as sociologists, should predict that there will constantly be conflicts over values and vested interests. Due to these conflicts, we should predict that there will be compromises. Because no political orientation, such as conservative or liberal, will typically have complete sway over the social policymaking process, we will see elements of both conservative and liberal ideas contained within the compromises that eventually become social policies. Radical social policy, on the other hand, will not, by and large, be accepted because radical groups will not have the resources or public backing to get support for their views. This is not to say that their proposed social policies would not be the best solutions to solve the problems. It just means that they will usually not have the power to influence Congress, the White House, or the general public to get their policies accepted.

Loss of Control Over the Social Problem

With our sociological perspective, we will observe that the group or organization that made the initial complaint and increased our consciousness about a social problem will many times tend to lose control over how the social problem is defined and what should be done about it as local, state, and federal governments take over responsibility for solving the problem. Probably, most groups that want a certain social condition to be seen as a social problem will not realize that, as they are successful in convincing others (especially governmental officials) of their views, they will lose control over the process they started. Political authorities, with more resources of money, power, prestige, and personnel, will tend to become the key players in further legitimizing a social condition as a social problem and in deciding what should be done and how it should be done.

Attention and Resources

We should also observe that as more attention and more resources are brought to bear on a social problem, the more likely the social problem will be solved or at least ameliorated. However, even after much attention is given and many resources are allocated, we still might not solve the problem because the social policy that we created is the wrong policy to solve that problem or because other existing social policies might work at cross-purposes with the new social policy, for example, as we discussed in Chapter 9 on drugs as to how a criminal model and medical model of social policies will mesh. Consequently, we should be conscious of the fact that although much attention to the problem and sufficient resources are needed, the right social policy is also needed.

Someone Will Be Dissatisfied

We should also predict that someone will always be dissatisfied with nearly any social policy we implement. There are a number of reasons for this. First, the problem and its potential solution can get redefined throughout the process of the social problem. With these different definitions, different groups with different definitions will talk at cross-purposes, resulting in increased dissatisfaction. Dissatisfaction will also arise because different groups will have different values and vested interests that will conflict. For example, conservative fundamentalists and liberals will have a difficult time in agreeing on abortion, gay marriage, stem cell research, and so on. Many rich people will not want to be taxed more to provide child care subsidies for single mothers and student loans for low-income students; they would rather be taxed less so that they can retain or even increase their wealth. As a consequence, we, as sociologists, should predict that there will be discontent no matter what social policy we try and no matter how effective it is.

Moreover, if anything goes wrong with the implementation of a social policy, we should predict that there will be criticism such as, "See, I told you so; that social policy will never solve this problem!" In other words, built into the process of solving a social problem will be the potential for dissatisfaction and criticism of social policies. Such a conclusion is not very consoling to political officials who need to face the wrath of different groups with different values, vested interests, and agendas, but at least they can take some solace in the fact that facing this dissatisfaction and criticism is part of the nature of social problem solving and is not necessarily due to their personal inadequacies.

Compromise

We should also predict that there almost always will be compromise between conservatives and liberals. As such, the social structure will be slower to change than liberals want and quicker to change than conservatives want, hence making both political groups somewhat dissatisfied. On the other hand, radicals will be totally dissatisfied with whatever social policy is developed because their own version of how to solve the social problem will be ignored and not given a chance. They will therefore lament that the social structure is changing at a snail's pace and conclude, "We will never solve our social problems!" In their frustration, we should observe that they may attempt to carry out some type of partial solution such as create communes, develop their own educational systems, seek new lifestyles through different religions and ways of viewing life, or move to other countries.

With such political freedom in our country where fundamentalist conservative, moderate conservative, liberal, and radical political orientations exist, there is the built-in potential for disagreement, dissatisfaction, opposition, and conflict. Consequently, we, as Americans, will need to live with a certain level of turmoil. It seems that this is the price we must pay in exchange for living in some semblance of a democratic society.[1]

Becoming Accustomed to Social Change

Given all the social change in the United States since the 1960s, such as the changes in civil rights for African Americans and the changes in laws and informal norms for women, homosexuals, the elderly, and the handicapped, giving these groups more opportunities than they ever had before, we have become more accustomed to social change as a part of our daily lives in this country.[2] Because we have experienced and become accustomed to more social change, we may be more prepared than were our ancestors to make more changes that will help us to solve our social problems.

We may also find that our experiencing so much social change will cause us to pick up the pace in attempting to solve social problems and be more willing to try a greater variety of social policies than we have tried in the past. Picking up the pace and trying a greater variety of social policies could very well result in our solving social problems at a faster rate. The problem that works against this faster rate, however, is the fact that there are millions more Americans now than there were a few generations ago. Hence, we will need more resources and more planning to solve our social problems of today.

Because there has been an increase in opportunities among various minorities and the continued emphasis on getting more education, we should observe a greater number and variety of people in our country who will have the chance to reflect on the social problems of our society. With more reflection by people from a variety of social and economic backgrounds, resulting in new ways of looking at old social problems, I cannot help but think that we will develop new and creative ideas to solve our social problems.

Capitalism and Social Problems

What can we say about the future of solving our social problems in relation to capitalism? First, as we discussed in Chapter 1, we should predict that capitalism will be around for a number of years to come because a substantial number of Americans have a good standard of living as a result of living in a capitalistic society, so it will continue to be given legitimacy as the best economic system so long as the majority of Americans perceive that they benefit from it. Consequently, for the time being, whatever will be done to solve social problems will need to be done within the context of capitalism.

Yet, ironically, as we have seen periodically throughout this book, it is capitalism that is a major cause of many of our social problems. For example, capitalism causes much inequality. It causes unemployment, leading to poverty and homelessness. It causes crime due to people not having jobs or having jobs that pay below the poverty line. Corporations illegally dispose of waste to make more profit. Some corporations (or their executive officers) steal retirement funds from their employees to make more profit. Corporations move plants to other countries to make more profit and leave workers and communities behind to fend for themselves. Corporations cut back on retirement and health benefits to make more profit (Heilbroner, 1991). As a result of these aspects of capitalism, millions of Americans are left out of the abundance that capitalism provides for many Americans. Knowing this influence, we, as sociologists, will continually need to study the influence of capitalism on social problems as a way to help us understand how to solve our social problems.

Environment, Population, and Standard of Living

Another problem that we, as Americans and as citizens of the world, will need to face is the problem of our environment in light of people wanting an ever higher material standard of living and in light of our rising world population. Just trying to meet the basic needs of food, shelter, and clothing for more people in the world will cause societies to produce more material goods, resulting in more pollution and more depletion of resources. Given these conditions, we, as sociologists, should predict that the problems of pollution and depletion of resources will continue in the near future and will be a daunting challenge for people living in the 21st century.

As a partial solution to this situation, Heilbroner (1991) called for providing more services for people instead of providing more material goods (p. 103). The problem with this solution is that the combination of increasing world population and people's desire for a higher material standard of living may be too strong for services alone to act as a satisfactory replacement for material goods. Possibly, if the population problem can be controlled, and if people are socialized to want fewer material goods, providing services could act as a substitute for material goods. For the foreseeable future, however, we should predict that the people of the world will continue to want more material goods, with the result that we will continue to pollute a lot, deplete a lot, and pile up more waste.

If the preceding thoughts hold true, will we someday be forced to alter our values, our lifestyles, and even our economic system? I do not know the answer to this question, but it seems that some things will eventually need to change. Someday, we or our future generations might have no choice but to change our way of living—if we want to survive as a species.

If we need to change, what kind of change might we see? One kind of change that continues nearly daily is technological change. Certain technologies will no doubt solve or ameliorate some social problems. For example, we might find ways to increase our use of mass transportation or discover ways of continuing to use cars and yet decrease pollution and the depletion of resources. Even when these kinds of technological discoveries are made, we, as sociologists, should predict that these technologies will not be implemented immediately because of the vested interests of those making a profit in the car and oil industries as they now exist. For example, if we stopped using oil all at once, oil companies and those industries affiliated with the oil companies would lose profit and workers in these companies would lose jobs.

We should therefore predict that powerful vested interests will fight to stop or slow down the implementation of any new technology that would solve or greatly ameliorate the social problems of pollution and the depletion of resources if that technology meant the loss of profit. So, although we might discover new technologies that could help to solve certain social problems, we should predict that other factors, such as power and vested interests, could stop or impede the use of these technologies. Hence, we should predict that technology alone will never be the complete answer to the solving of social problems because the problem of vested interests will also need to be overcome.

Building a Sustainable Society

Brown (1987), in his book *In Building a Sustainable Society,* suggested that we might need to change from a more competitive, growth-oriented, oil-based society to a more cooperative, population-stable, energy-independent type of society. This new way of living is more likely to happen in the future as we realize that we might need to change to survive. In the meantime, we should predict that there will be huge conflicts over competing vested interests and values that will slow down the change from our current way of living to a more sustainable way of living, where we cooperate more within and among societies, we have a population-stable world, and we have an energy-independent society and world.

Someday, we may indeed move to a more sustainable type of society and world, but not without considerable turmoil. This is so because although everyone wants to survive, they do not want to change their values, hurt their vested interests, and decrease their standard of living. We, as sociologists, should therefore predict that as we move toward a more sustainable society, it will be an arduous process. In other words, our short-term interests of wanting to maintain a certain way of life will conflict with our long-term interests of wanting to survive as a species.

I imagine that what I just stated may give you a sense of pessimism about the solving of our social problems. I do not mean or want to be pessimistic, but as a sociologist I need to look at what I think will happen, not what I want to happen. Yes, personally, I hope that all social problems will be solved—and solved sooner rather than later. But as a sociologist, I must look objectively at the factors that increase or decrease the likelihood of solving our social problems. And when I do that, I cannot help but conclude that there are strong forces such as power, values, and vested interests that will slow down the process of solving our social problems.

Given these strong forces, probably what we will see during the 21st century will be the changing of our society and the world in a disjointed way—sometimes fast and sometimes slow, sometimes organized and planned way and sometimes chaotic and disorganized. Over time, social problems will be solved or greatly ameliorated because more people and leaders will realize the need for solving them and will want them solved, but this trend will not occur overnight just because we wish these social problems were solved. Wishing, however, is at least a step in that direction. Awareness, education, organizing, and taking action are other key steps.

Conclusion

Life Is Not Always Fair

As you can see, solving a social problem is not easy. African people, from the moment they were put in chains more than 300 years ago, probably thought, "How can we regain our freedom?" Women during the 19th and 20th centuries questioned why they could not vote or hold political office or why they could not become presidents of companies, ministers, or doctors. Poor people of each generation wished they had money and material comforts like nonpoor people had. Yet possessing the desires for more freedom, equality, and money did not immediately change the social conditions of these people.

Hundreds of years would pass before the descendants of Africans gained some semblance of freedom. Only during the past 30 to 40 years of human history have women been able to participate much more fully in societies. Poverty, even in the America of today, can continue throughout a person's life. So, wanting something to change and feeling deep in one's heart that it should change does not mean it will change. Many Americans and people in other parts of the world who face, and have faced, various social problems have no doubt wondered and lamented the following:

> Why do we have to go through this? We are not bad people. We are good people. We work hard. We are honest. We treat people fairly and with respect. Why are we treated this way? We are all humans. Why can't we all treat each other as fellow humans?

Whether we have been a part of a larger social problem that has hurt us or whether we have experienced personal turmoil (e.g., our parents went through a divorce and the whole process hurt us deeply), each of us has wondered, "This should not happen to me. I did not do anything wrong. I should not need to endure these things!"

This happens to many people throughout the world who are living in dire poverty and do not know where their next meal will come from or how they can make enough money to buy food to survive. This was made clear to me personally when I visited the island country of Haiti and saw immense poverty and extreme deprivation. Everywhere I looked, I saw people with almost nothing and I thought, "How do they survive?" A Haitian woman sits all day on the side of the road in 90 to 100 degree temperatures hoping to sell some used pairs of shoes. A Haitian man puts a mat down on the edge of a busy street and tries to sell some fruit. Someone else works to sell a few pieces of clothing. Many Haitians will start at 4 o'clock in the morning to walk and then ride from their rural mountain homes to get to the city to sell the goods they have to offer and then make the 2- to 4-hour trek back home during the late afternoon and early evening. The same thing is done the next day, and the next day, and so on.

Life is not fair. In fact, life is not fair for many people currently living in many parts of the world, and life was not fair for many people throughout history. Not all of us begin our lives with equal opportunities; sufficient food, shelter, and clothing; and freedom from fear and want. For many people, life starts out offering little or no dignity, respect, or means to survive. As we know, life can be cruel to people even when they had nothing to do with their situation. They just happened to be born poor in a poor country. They just happened to be born to a group of people that has been discriminated against and therefore has faced blocked opportunities—possibly for generations. They were not the cause of their situation. It existed before they were born. Yet they will need to endure their situation all of their lives. They were born. They became little boys or little girls and were naive and innocent and did not understand their situation. But as they lived more years, they began to realize that they were in a never-ending situation—and not one of their making. For most people in these kinds of social conditions, their life situation did not change at all throughout their lives—for example, slaves during the 1700s wanting their freedom, women during the 1800s wanting to vote, and poor people in the inner-city ghettoes of the 1970s wanting good-paying factory jobs when those jobs were disappearing to Mexico, South Korea, and elsewhere.

Use the Sociological Perspective to Understand More Clearly

As we use a sociological perspective to understand the nature of social problems, what becomes abundantly clear is that the larger social structure—the laws, the informal norms, the ideologies, the statuses and roles, the economy, the political system, the inequality of power and money, and the values and vested interests that people have—are usually too powerful for one person or a few people to change. The social structure, in a sense, "swallows up" any attempt at change regarding objectives such as having equal opportunity, meeting basic needs, being treated with dignity and respect, and having some amount of redistribution of resources so that people can live decently.

Once something has become a part of the social structure, it is hard to change. As Emile Durkheim, the great French sociologist, pointed out, the social structure is external to us and yet coercive on us (Durkheim, 1966). That is, once we, as humans, create a social structure, it controls or highly influences us, for example, our thoughts, feelings, beliefs, and actions. It is hard to change the social structure because many people become accustomed to living a certain way of life and because people have vested interests of money, power, and prestige to protect. So long as a certain social structure continues, social problems and the personal problems stemming from this social structure will continue. When we begin to realize the relationship among the social structure, social problems, and personal troubles, we have developed what Mills (2000) called a sociological imagination.

Change Parts of the Social Structure

One important conclusion we can make is that if we want to solve or ameliorate our social problems, we will need to change various parts of the social structure. Those parts could be values, beliefs, laws, informal norms, or some aspects of the economy. With regard to most social problems, we will probably need to change a number of parts to solve these problems.

Reflecting on the civil rights movement and the women's rights movement, for example, we can say that social change, the decline of a social problem, and the improvement of people's personal lives can indeed occur (see the last parts of our causal model in Chapter 1). Recent history also shows that various parts of the social structure need to change if improvement in the social problem is to occur. For example, laws might need to change so that certain categories of people can have opportunity. Once laws are passed, people can resort to the court system to redress wrongs. As a consequence, the court system can be a powerful tool to right the wrongs that people of the past created via the social structures they created. Beliefs might need to change. For example, if we begin to believe that all humans should have equal opportunity, we can study and become more aware of the parts of the social structure that hinder this new belief. Such consciousness will lead people to come together to form groups, organizations, and social movements to achieve greater equal opportunity (again, refer to our theory in Chapter 1). Informal norms might need to change. In the past, the informal norms may have called for discrimination of certain categories of people. Once the informal norms change, discrimination decreases, allowing more equal opportunity. If the economy is not meeting the survival needs of people, parts of the economy might need to change or the government might need to step in to meet people's survival needs that the economy by itself is unable to meet.

Understand the Social Construction of Reality

There is one more point I want you to be aware of about the social structure. We, as humans, created the social structures that have caused the social problems we have today. For example, the early colonists socially constructed the social structure of slavery in colonial America and the legal and informal discrimination of the 1800s and 1900s that resulted in African Americans being at a substantial disadvantage in owning land, starting their own businesses, going to school, going to college, going to graduate school to have the choices of many careers, and having the same opportunities as whites to be upwardly mobile, to live anywhere they want and can afford to live, and to vote and hold local, state, and national political offices. Early colonists socially constructed social structures that prohibited women from voting, holding political office, owning land, starting businesses, going to college, having all kinds of jobs and careers, and being able to have the same overall opportunities as did men. Since the 1960s, business and industrial owners and managers have socially reconstructed the economic part of our social structure in terms of replacing workers with robots, creating more service-producing industries than goods-producing industries, creating a number of high-wage service jobs and many low-wage service jobs, and moving many factory jobs to other countries, resulting in high unemployment rates for people in inner cities with the resulting perpetuation of poverty, homelessness, higher crime rates, and psychological problems such as stress, low self-esteem, and depression.

Plan for New Social Construction of Reality

In other words, we, as humans, have socially constructed social structures throughout history without realizing their negative consequences (Merton, 1967). We have created social structures without realizing the havoc they wreaked on society in general or on certain groups of people in particular. Moreover, many groups of people can experience this havoc for hundreds of years. Given that this has happened throughout much of human history, it makes sense for us to (a) plan the kind of social structures we want given certain criteria that all or nearly all of us could agree on as guides for our future societies and (b) study how new social structures could have more positive consequences and fewer negative consequences for humans.

As for specific criteria we may wish to use in creating future social structures, the following are some we might consider using. What social structures are just? What social structures are humane? What social structures give people equal or near equal opportunity to get ahead? What social structures help people to meet their material needs? What social structures give people considerable freedom? Given these criteria, or whatever criteria the people of a society or world would choose to go by, we could set about to plan and create social structures that satisfy such criteria.

We are at the point in human history where we now realize that we need to be more conscious of the social structures we create. There are too many negative consequences that can happen to people when social structures are created without forethought. Now, more than at any other time in history, we realize that we, as humans, have created our social structures and that we can change them.[3] More than ever before in human history, we realize we can have more control over the social structures we create, the consequences these social structures bring about, and thus the way we live our lives. More and more people agree that we have it within our power to create new social structures that will help all of us to live more fulfilling lives.

To create social structures based on the criteria we create, we will need to have (a) much research in the area of planning, (b) much democratic discussion to get all sides of an issue heard, and (c) much testing of our plans before we create new social structures.

We will make mistakes. Although we will try to think of all the positive and negative consequences that may result due to changing our social structures, we will probably not uncover all of the consequences before we create new social structures. We will need to accept that we are not perfect and therefore do not know all and see all. Yet we know that we have past history and other contemporary societies to study and enlighten us as well as recent research to rely on to help us uncover many consequences. So, we already have existing wisdom on which to draw.

Realize Hope for the Future

In summary, more than ever before in history, we realize the influence that social structures have on humans. We realize that we created these social structures and that we can change them. We realize that we can create criteria that can act as guides as we plan new social structures to meet human needs. We realize that we have past history and recent research findings to help us plan. We realize that we can test our plans first to "work out the bugs." So, there are many things we realize that can help us to socially construct future social structures that will solve our social problems.

There is no doubt that social problems are not easy to solve. Yet more than ever before, we understand more fully the workings of society in general and of social problems in particular. We know how variables relate to each other and what variables cause other variables to change. We realize that we created the social structures within which we live and that we can change these social structures.

Most of us want to solve our social problems and live in a society (and world) that is humane and just. Knowing that we have already made progress in ameliorating a number of our social problems (e.g., decreasing poverty among elderly people, decreasing prejudice and discrimination against various minorities, increasing opportunities for more people), I cannot help but conclude that now, more than ever before in history, we can and should have hope that we will someday solve our social problems.

Questions for Discussion

1. Do you think we will someday create a sustainable society? If so, how?

2. What do you predict for the future of solving our social problems?

3. Will we ever solve our current social problems? What is your reasoning?

4. How can we solve our social problems in the context of capitalism?

5. What do you think will be the key independent variables during the 21st century that will help us to solve our social problems?

6. What do you think will be the key barriers to the solving of our social problems during the 21st century?

7. Which social problems do you think will be solved, and which ones do you think will not be solved, during the next 20 to 30 years? What is your reasoning?

8. In the future, where will the government fit into the solving of social problems?

9. Do you think that someday we, as humans, will share a common set of values that will help us to solve our social problems?

10. Where should we, as a society, go from here to solve our social problems?

CHAPTER 1

Notes

1. For additional ideas on the subjective and objective elements of a social problem, see Blumer's (1971) journal article.

2. You might compare my definition with the definitions of other people in the field, including the following. "A social problem is a condition caused by factors built into the social structure of a particular society that systematically disadvantages or harms a specific segment or a significant number of the society's population" (Curran & Renzetti, 2000, p. 3). "A social problem is a social condition that a segment of society views as harmful to members of society and in need of remedy" (Mooney, Knox, & Schact, 2002, p. 3). "When most people in a society agree that a condition exists that threatens the quality of their lives and their most cherished values, and they also agree that something should be done to remedy that condition" (Kornblum & Julian, 2001, p. 4). "Social problems are issues that substantial numbers of the society view as violations of society's social norms or expectations, and about which people believe something can and should be done" (Palen, 2001, p. 10). Parrillo (2002), in his *Contemporary Social Problems*, stated, "We also must know that recognized social problems have the following four components: 1. They cause physical or mental damage to individuals or society; 2. They offend the values or standards of some powerful segment of society; 3. They persist for an extended period of time; and 4. They generate competing proposed solutions because of varying evaluations from groups in different social positions within society, which delays reaching consensus on how to attack the problem" (pp. 4–5). Eitzen and Zinn (2000), in their *Social Problems,* stated that there are "two main types of social problems: (1) acts and conditions that violate the norms and values present in society, and (2) societally induced conditions that cause psychic and material suffering for any segment of the population" (p. 7).

3. For excellent discussions of Comte's life and how he laid the foundation for the beginning of sociology, see Ritzer (1996a) and Turner, Beeghley, and Powers (1995a, 1995e).

4. For excellent discussions of Durkheim's life and work and of how he began the formal discipline of sociology in the academic setting, see Ritzer (1996b) and Turner, Beeghley, and Powers (1995b, 1995f).

5. For excellent discussions of the life and work of Marx, see Ritzer (1996c) and Turner, Beeghley, and Powers (1995c, 1995g).

6. For an excellent discussion on the life and work of Weber, who was the major figure in starting the formal discipline of sociology in Germany during the late 1800s and early 1900s, see Ritzer (1996d) and Turner, Beeghley, and Powers (1995d, 1995h).

7. The sociologists who take this stance use the writings of Weber, the famous German sociologist, who asserted that sociologists need to remain as objective as possible so long as we are "wearing the hat" of a sociologist. When we are not in our role as sociologists, such as citizens casting our votes or belonging to certain political parties with certain stances on issues, we can and should voice our opinions. See Weber's (1946) discussion on this topic titled "Science as a Vocation." For example, he asserted, "Politics is out of place in the lecture-room. . . . To take a practical political stand is one thing, and to analyze political structures and party positions is another. When speaking in a political meeting about democracy, one does not hide one's personal standpoint; indeed, to come out clearly and take a stand is one's damned duty. The words one uses in such a meeting are not means of scientific analysis but means of canvassing votes and winning over others. They are not plowshares to loosen the soil of contemplative thought; they are swords against the enemies: such words are weapons. It would be an outrage, however, to use words in this fashion in a lecture or in the lecture-room. . . . Whenever the man of science introduces his personal value judgment, a full understanding of the facts ceases" (pp. 145–146). See also Weber's (1949) discussion titled "'Objectivity' in Social Science and Social Policy," where he maintained that "an empirical science cannot tell anyone what he should do—but rather what he can do" (p. 54).

8. For a more detailed discussion on this matter, see Manis's (1974) journal article.

9. For a more detailed explanation, see Turner's (1998) *The Structure of Sociological Theory*.

10. See a famous book in sociology by Berger and Luckmann's (1967) *The Social Construction of Reality*. In their book, they did an excellent job of showing how humans create all kinds of social phenomena.

11. See Weber (1968), where he discussed three dimensions of inequality: class or money, party or power, and status or prestige. Sociologists use these three dimensions to measure inequality.

12. See Turner (1998, p. 158), where he reformulated the ideas of Weber in a propositional format. See also Weber (1968). Turner noted that Weber suggests that if society socializes people to be upwardly mobile and tells them that they have equal opportunity when in fact they do not have equal opportunity, we have created social conditions for these people to question the legitimacy of the existing social conditions.

13. For others who have suggested that there may be a sequence or phases or stages of social problems, see Blumer (1971); Case (1924); Frank (1925); Fuller and Myers (1941); Peyrot (1984); and Spector and Kitsuse (1973).

14. For various theoretical propositions based on Blau's work that I use at various points throughout this book, see Turner (2003).

15. A good place to start to think about what will happen to capitalism and any indicators as to when is to consider the views of Schumpeter (1942).

References

Berger, P. L., & Luckmann, T. (1967). *The social construction of reality*. Garden City, NY: Anchor Books.

Blau, P. M. (1964). *Exchange and power in social life*. New York: John Wiley.

Blumer, H. (1971). Social problems as collective behavior. *Social Problems, 18,* 298–305.

Case, C. M. (1924). What is a social problem? *Journal of Applied Sociology, 8,* 268–273.

Cooley, C. H. (1902). *Human nature and the social order*. New York: Schocken Books.

Curran, D. J., & Renzetti, C. M. (2000). *Social problems: Society in crisis* (5th ed.). Boston: Allyn & Bacon.

Durkheim, E. (1938). What is a social fact. In *The rules of sociological method* (pp. 1–13). New York: Free Press.

Eitzen, D. S., & Zinn, M. B. (2000). *Social problems* (8th ed.). Boston: Allyn & Bacon.

Frank, L. K. (1925). Social problems. *American Journal of Sociology, 30,* 462–473.

Fuller, R. C., & Myers, R. R. (1941). The natural history of a social problem. *American Sociological Review, 6,* 320–328.

Gouldner, A. W. (1960). The norm of reciprocity. *American Sociological Review, 25,* 161–178.

Kornblum, W., & Julian, J. (2001). *Social problems* (10th ed.). Upper Saddle River, NJ: Prentice Hall.

Manis, J. G. (1974). Assessing the seriousness of social problems. *Social Problems, 22,* 1–15.

Marx, K. (1964). Estranged labor. In *The economic and philosophic manuscripts of 1844* (pp. 106–119). New York: International Publishers.

Marx, K. (1972). Theses on Feuerbach. In R. C. Tucker (Ed.), *The Marx–Engels reader*. New York: Norton.

Merton, R. K. (1938). Social structure and anomie. *American Sociological Review, 3,* 672–682.

Merton, R. K. (1967). Manifest and latent functions. In *On theoretical sociology* (pp. 73–138). New York: Free Press.

Merton, R. K. (1968a). Continuities in the theory of reference groups and social structure. In *Social theory and social structure* (enlarged ed., pp. 335–440). New York: Free Press.

Merton, R. K., with Rossi, A. S. (1968b). Contributions to the theory of reference group behavior. In *Social theory and social structure* (enlarged ed., pp. 279–334). New York: Free Press.

Mills, C. W. (1959). *The sociological imagination*. London: Oxford University Press.

Mooney, L., Knox, D., & Schact, C. (2002). *Understanding social problems*. Belmont, CA: Wadsworth/Thomson Learning.

Palen, J. J. (2001). *Social problems for the twenty-first century*. Boston: McGraw–Hill.

Parrillo, V. N. (2002). *Contemporary social problems* (5th ed.). Boston: Allyn & Bacon.

Parsons, T. (1951). *The social system.* New York: Free Press.

Peyrot, M. (1984). Cycles of social problem development: The case of drug abuse. *Sociological Quarterly, 25,* 83–95.

Ritzer, G. (1996a). Auguste Comte. In *Classical sociological theory* (2nd ed., pp. 87–113). New York: McGraw–Hill.

Ritzer, G. (1996b). Emile Durkheim. In *Classical sociological theory* (2nd ed., pp. 183–216). New York: McGraw–Hill.

Ritzer, G. (1996c). Karl Marx. In *Classical sociological theory* (2nd ed., pp. 149–182). New York: McGraw–Hill.

Ritzer, G. (1996d). Max Weber. In *Classical sociological theory* (2nd ed., pp. 217–263). New York: McGraw–Hill.

Schumpeter, J. A. (1942). *Capitalism, socialism, and democracy.* New York: Harper & Row.

Spector, M., & Kitsuse, J. I. (1973). Social problems: A re-formulation. *Social Problems, 21,* 145–159.

Sutherland, E. H. (1940). White-collar criminality. *American Sociological Review, 5,* 1–12.

Turner, J. H. (1991). *The structure of sociological theory* (5th ed.). Belmont, CA: Wadsworth.

Turner, J. H. (1998). *The structure of sociological theory* (6th ed.). Belmont, CA: Wadsworth.

Turner, J. H. (2003). Dialectical exchange theory: Peter M. Blau. In *The structure of sociological theory* (7th ed., pp. 294–307). Belmont, CA: Wadsworth/Thomson Learning.

Turner, J. H., Beeghley, L., & Powers, C. H. (1995a). The origin and context of Auguste Comte's thought. In *The emergence of sociological theory* (3rd ed., pp. 13–28). Belmont, CA: Wadsworth.

Turner, J. H., Beeghley, L., & Powers, C. H. (1995b). The origin and context of Emile Durkheim's thought. In *The emergence of sociological theory* (3rd ed., pp. 284–309). Belmont, CA: Wadsworth.

Turner, J. H., Beeghley, L., & Powers, C. H. (1995c). The origin and context of Karl Marx's thought. In *The emergence of sociological theory* (3rd ed., pp. 102–128). Belmont, CA: Wadsworth.

Turner, J. H., Beeghley, L., & Powers, C. H. (1995d). The origin and context of Max Weber's thought. In *The emergence of sociological theory* (3rd ed., pp. 168–189). Belmont, CA: Wadsworth.

Turner, J. H., Beeghley, L., & Powers, C. H. (1995e). The sociology of Auguste Comte. In *The emergence of sociological theory* (3rd ed., pp. 29–46). Belmont, CA: Wadsworth.

Turner, J. H., Beeghley, L., & Powers, C. H. (1995f). The sociology of Emile Durkheim. In *The emergence of sociological theory* (3rd ed., pp. 310–350). Belmont, CA: Wadsworth.

Turner, J. H., Beeghley, L., & Powers, C. H. (1995g). The sociology of Karl Marx. In *The emergence of sociological theory* (3rd ed., pp. 129–167). Belmont, CA: Wadsworth.

Turner, J. H., Beeghley, L., & Powers, C. H. (1995h). The sociology of Max Weber. In *The emergence of sociological theory* (3rd ed., pp. 190–232). Belmont, CA: Wadsworth.

Weber, M. (1946). Science as a vocation. In H. H. Gerth & C. W. Mills (Eds.), *From Max Weber: Essays in sociology* (pp. 129–156). New York: Oxford University Press.

Weber, M. (1949). "Objectivity" in social science and social policy. In *The methodology of the social sciences* (pp. 49–112). New York: Free Press.

Weber, M. (1968). The distribution of power within the political community: Class, status, party. In *Economy and society: An outline of interpretive sociology* (pp. 926–940). New York: Bedminster.

CHAPTER 2

Notes

1. Ryan (1981) discussed the need for people to have access to resources such as medical care, quality education, and housing, resulting in income inequality not having such a dire effect on people with low incomes.

2. For a succinct overview of libertarian, conservative, and liberal ("social democratic") philosophies, see Fine and Shulman's (2003, pp. 6–13) *Talking Sociology*.

3. See, for example, the newspaper article by King (2005). This article discusses the continuing debate about whether or not creationism, or a more recent version of creationism called *intelligent design*, should be taught in a science classroom or in any classroom in a public school.

4. This principle can also be applied to other countries with congresses and parliaments.

5. Broder (2002) discussed how decreasing tax cuts at the federal level was resulting in less aid for state governments that in turn needed to decrease state funding such as Medicaid for the poor and cut aid to local governments. He noted that the federal tax cuts, besides decreasing federal surpluses and increasing federal deficits, meant that "state and local governments are raising taxes and slashing vital services in order to balance their budgets" (p. A11). Hence, the shortsightedness of decreasing taxes has numerous negative effects on state and local governments and on families, especially lower income ones.

6. As Schumpeter (1976) noted, "The first and foremost aim of each political party is to prevail over the others in order to get into power or to stay in it" (p. 279).

7. For a discussion of American values, see Williams's (1970) *American Society*.

8. Eitzen and Sage (1997) noted that the University of Tennessee of the Southeastern Conference broke the racial barrier when it signed an African American football player for the 1966–1967 school year.

9. When I was an undergraduate at the University of Tennessee during the early 1960s, African Americans did not live on campus, were not in the fraternities and sororities, and were not athletes for the university. The few black students who were attending the university at the time lived in African American homes. They would take the bus across town in Knoxville, attend classes, and return to the private homes during the evening. As a result, the African American students were less likely to participate in extracurricular activities than were the white students who lived in the dorms and had much more access to extracurricular activities.

10. Max Weber, the great German sociologist, helped us to realize the potential role that values can play in social action in his book, *The Protestant Ethic and the Spirit of Capitalism* (Weber, 1958).

11. See, for example, Blau's (1964) *Exchange and Power in Social Life*. A key point that exchange theory makes is that not just financial relationships are exchange relationships and that nearly all social relationships have some form of exchange to them. Turner (1991) converted Blau's ideas into interrelated propositions that can be tested empirically for their validity. For example, one of Turner's propositions from Blau is the following: "The more profit people expect from one another in emitting a particular activity, the more likely they are to emit that activity" (p. 331).

References

Berger, P. L., & Kellner, H. (1981). *Sociology reinterpreted: An essay on method and vocation.* Garden City, NY: Anchor Books.

Blau, P. M. (1964). *Exchange and power in social life.* New York: John Wiley.

Blumer, H. (1971). Social problems as collective behavior. *Social Problems, 18,* 298–305.

Broder, D. S. (2002, July 31). The states' dilemma. *Louisville Courier–Journal,* p. A11.

Durkheim, E. (1933). *The division of labor in society* (G. Simpson, Trans.). New York: Free Press.

Eitzen, D. S., & Leedham, C. S. (1998). *Solutions to social problems: Lessons from other societies.* Boston: Allyn & Bacon.

Eitzen, D. S., & Leedham, C. S. (2001). *Solutions to social problems: Lessons from other societies* (2nd ed.). Boston: Allyn & Bacon.

Eitzen, D. S., & Sage, G. H. (1997). *Sociology of North American sport* (6th ed.). Madison, WI: Brown & Benchmark.

Fine, G. A., & Shulman, D. (2003). *Talking sociology* (5th ed.). Boston: Allyn & Bacon.

Kerbo, H. R. (2006). *Social stratification and inequality: Class conflict in historical, comparative, and global perspective* (6th ed.). Boston: McGraw–Hill.

King, R. (2005, August 22). Evolution debate is playing out in Hoosier schools. *Louisville Courier Journal*, pp. B1–B2.

Kingdon, J. W. (1993). How do issues get on public policy agendas? In W. J. Wilson (Ed.), *Sociology and the public agenda* (pp. 40–50). Newbury Park, CA: Sage.

Lenski, G. E. (1984). *Power and privilege: A theory of social stratification.* Chapel Hill: University of North Carolina Press.

Marx, K., & Engels, F. (1978). *The Communist manifesto.* Chicago: Charles H. Kerr.

Mills, C. W. (1956). *The power elite.* London: Oxford University Press.

Peyrot, M. (1984). Cycles of social problem development: The case of drug abuse. *Sociological Quarterly, 25,* 83–95.

Piven, F. F., & Cloward, R. A. (1971). *Regulating the poor: The functions of public welfare.* New York: Vintage.

Ryan, W. (1981). *Equality.* New York: Vintage.

Schoenfeld, A. C., Meier, R. F., & Griffin, R. J. (1979). Constructing a social problem: The press and the environment. *Social Problems, 27,* 38–61.

Schumpeter, J. A. (1976). *Capitalism, socialism, and democracy.* New York: Harper & Row.

Turner, J. H. (1991). *The structure of sociological theory* (5th ed.). Belmont, CA: Wadsworth.

Weber, M. (1958). *The Protestant ethic and the spirit of capitalism.* New York: Scribner.

Weiss, C. H. (1993). The interaction of the sociological agenda and public policy. In W. J. Wilson (Ed.), *Sociology and the public agenda* (pp. 23–39). Newbury Park, CA: Sage.

Williams, R. M., Jr. (1970). *American society: A sociological interpretation* (3rd ed.). New York: Alfred A. Knopf.

Wilson, W. J. (1993). Can sociology play a greater role in shaping the national agenda? In W. J. Wilson (Ed.), *Sociology and the public agenda* (pp. 3–22). Newbury Park, CA: Sage.

CHAPTER 3

Notes

1. For a further discussion of these three dimensions of inequality, see Weber's (1968, pp. 926–940) *Economy and Society.*

2. Marx (1967), in *Capital: A Critique of Political Economy,* noted how those who have money in a capitalistic society will also have more power.

3. For a discussion of who holds certain positions and therefore has more power in our society, see Mills's (1959a) *The Power Elite.*

4. For an excellent discussion of people of the corporate class and how they are top officers in one large corporation and serve on the boards of other large corporations, see Kerbo's (2003) *Social Stratification and Inequality.*

5. For a revealing discussion of this process, see Barlett and Steele's (1992) *America: What Went Wrong?*

6. Marx and Engels (1992) put it this way: "The ruling ideas of each age have ever been the ideas of its ruling class" (p. 40). In an updated variation on this same theme, see Bookman's (2002) newspaper article.

7. For an excellent discussion of Weber's ideas on legitimacy and related consequences, see Turner, Beeghley, and Powers's (2002) *The Emergence of Sociological Theory.*

8. In one of my recent social problems classes, no one voted to increase inequality, 12.5% voted to maintain the current inequality, and 87.5% voted to decrease the inequality in our country. The breakdown of conservatives and liberals in the class was interesting. Among conservative students, 30% wanted to keep the current inequality, whereas 70% wanted to decrease it. Among liberal students, 100% wanted to decrease inequality. So, even though the liberals were more in favor of decreasing the current inequality than were conservatives, a strong majority of conservative students wanted to decrease inequality as well.

References

Barlett, D. L., & Steele, J. B. (1992). *America: What went wrong?* Kansas City, MO: Andrews & McMeel.

Berger, P. L., & Kellner, H. (1981). *Sociology reinterpreted: An essay on method and vocation.* Garden City, NY: Anchor Books.

Bookman, J. (2002, January 13). "Little guys"—you and I—squeezed out. *Louisville Courier-Journal*, p. D1.

Eitzen, D. S., & Zinn, M. B. (2003). *Social problems* (9th ed.). Boston: Allyn & Bacon.

Kerbo, H. R. (2003). *Social stratification and inequality: Class conflict in historical, comparative, and global perspective* (5th ed.). Boston: McGraw–Hill.

Krugman, P. (2006a, March 27). Looking at income disparity. *Louisville Courier-Journal*, p. A9.

Krugman, P. (2006b, March 1). Rise of the American oligarchy. *Louisville Courier-Journal*, p. A7.

Marx, K. (1967). *Capital: A critique of political economy* (Vol. 1). New York: International Publishers.

Marx, K., & Engels, F. (1992). *The Communist manifesto.* New York: Bantam.

Mills, C. W. (1959a). *The power elite.* London: Oxford University Press.

Mills, C. W. (1959b). *The sociological imagination.* London: Oxford University Press.

Turner, J. H., Beeghley, L., & Powers, C. H. (2002). *The emergence of sociological theory* (5th ed.). Belmont, CA: Wadsworth.

Weber, M. (1968). *Economy and society: An outline of interpretive sociology.* New York: Bedminster.

CHAPTER 4

Notes

1. Eitzen and Zinn (2003), in their book *Social Problems*, reported that 12.1 million American children (under 18 years of age) were poor as of 1999 (p. 184) and that, although the overall poverty rate in the United States was 11.8% in 1999, the poverty rate for children was 16.9% (p. 183). Children make up 40% of all the poor people in the United States (p. 210).

2. Yetter (2003) noted that mostly poor single mothers "are being forced to quit jobs or drop out of school, and some will wind up back on the welfare rolls," according to state officials and advocates who work with the poor (p. A1). Yetter noted that 25 states "are not able to serve all families who apply" (p. A4). Khrystal Johnson of Louisville, Kentucky, said, "I can't afford with my weekly paycheck to put two kids in day care" (p. A4). Johnson estimated that her day care costs are $230 per week for her two children. Also, an unintended consequence of the lack of funding, noted Yetter, is that it will "cause parents to place children in inadequate care or leave them alone" (p. A4).

3. See TaxCreditResources.org (n.d. a). Currently, 14 states plus the District of Columbia have state programs.

4. See TaxCreditResources.org (n.d. b), which reported that in 1999, the earned income tax credit lifted 4.8 million people above the poverty line.

5. We would need to discuss various ways to do this. For example, we could look into the possibility of issuing something like a credit card that, when used to pay the cashier, would indicate the person's most recent yearly income.

6. See U.S. Department of Labor (n.d. b). For a good overview of the program, see also Almanac of Policy Issues (n.d.).

7. In *The Structure of Sociological Theory*, Turner (1991) created a number of theoretical propositions from the ideas of Max Weber. The two propositions that are important for the preceding discussion are as follows: (a) "The lower the rates of mobility of social hierarchies of power, prestige, and wealth, the more intense the level of resentment among those denied opportunities and, hence, the more likely they are to withdraw legitimacy" (p. 198); and (b) "The greater the degree of withdrawal of legitimacy from political authority, the more likely is conflict between superordinates and subordinates" (p. 198).

8. For a succinct discussion on our housing crisis, see Eitzen and Zinn's (2003, pp. 157–162) *Social Problems*.

9. Eitzen and Zinn (2003, p. 159) noted that whereas as much as 40% of the housing in Germany, France, and The Netherlands is owned by the government, only 1.3% of U.S. housing is publicly owned.

References

Almanac of Policy Issues. (n.d.). *Unemployment compensation.* [Online.] Available: www.policyalmanac.org/social_welfare/archive/unemployment_compensation.shtml

Cornell Law School. (n.d.). *Unemployment compensation law: An overview.* [Online.] Available: www.law.cornell.edu/topics/unemploymentcompensation.html

Eitzen, D. S., & Zinn, M. B. (2003). *Social problems* (9th ed.). Boston: Allyn & Bacon.

Friedman, M. (1962). *Capitalism and freedom.* Chicago: University of Chicago Press.

Gans, H. J. (1995). *The war against the poor: The underclass and antipoverty policy.* New York: Basic Books.

Kerbo, H. (2006). *Social stratification and inequality: Class conflict in historical, comparative, and global perspective* (6th ed.). Boston: McGraw–Hill.

Palen, J. J. (1997). *The urban world* (5th ed.). New York: McGraw–Hill.

Pear, R. (1999, October 4). 44.3 million have no health insurance. *Louisville Courier–Journal,* p. A1.

Pear, R. (2003, June 28). Medicare reform still not certain: Compromise will not come easily. *Louisville Courier–Journal,* pp. A1, A4.

Sanders, B. (2000, Spring). The "booming" economy. *Sanders Scoop,* p. 3. (Sanders for Congress newsletter)

TaxCreditResources.org. (n.d. a). *State ETC programs.* [Online.] Available: www.taxcreditresources.org

TaxCreditResources.org. (n.d. b). *What is the earned income credit (EIC)?* [Online.] Available: www.taxcreditresources.org

Tax Policy Center. (n.d.). *Poverty levels, tax thresholds, and federal tax amounts for different family sizes with earnings equal to the poverty level, 1993–2009.* [Online.] Available: www.taxpolicycenter.org

Turner, J. H. (1991). *The structure of sociological theory* (5th ed.). Belmont, CA: Wadsworth.

Unemployment hits 9-year high. (2003, July 4). *Louisville Courier–Journal,* p. A1.

University of Texas. (n.d.). *Unemployment rate, Austin–Round Rock MSA, Texas, and U.S., 1999–2003.* [Online.] Available: www.utexas.edu/depts/bbr/austindex/snapshot/unemployment/unemprate.pdf

U.S. Department of Health and Human Services. (n.d.). *The 2003 HHS poverty guidelines.* [Online.] Available: http://aspe.hhs.gov/poverty/03poverty.htm

U.S. Department of Labor. (n.d. a). *History of federal minimum wage rates under the Fair Labor Standards Act, 1938–1996.* [Online.] Available: www.dol.gov/esa/minwage/chart.htm

U.S. Department of Labor. (n.d. b). *Unemployment insurance extended benefits.* [Online.] Available: http://workforcesecurity.doleta.gov/unemploy/extenben.asp

Wilson, W. J. (1997). *When work disappears: The world of the new urban poor.* New York: Alfred A. Knopf.

Yetter, D. (2003, June 23). Many needy parents denied child-care help: State's long waiting list may push some on welfare. *Louisville Courier–Journal,* pp. A1, A4.

CHAPTER 5

Notes

1. For three different political views on affirmative action, see Fine and Shulman (2003).

2. Kinsley (2003) stated that race will continue to be seen as a legitimate variable to be used to decide who gets into a school just as other variables, such as whether the applicant is an athlete, plays a musical instrument, is an artist, has parents who give a lot of money to the school, has parents who are alums, or has high grades, are used. Also, Thomas (2003) noted, "There is not a single word in the U.S. Constitution about educating citizens of the United States, nor is there a word about diversity as a 'compelling interest' of the government" (p. A7). He added that we need to place more emphasis on primary and secondary education in our country because this "would ensure that minority (and majority) kids learn their subjects and qualify for admission based on merit" (p. A7).

3. There are a number of reasons given to justify implementing and continuing affirmative action: (a) the 200 years of slavery and 100 years of blatant and severe prejudice, discrimination, and segregation; (b) the current institutional discrimination that puts minorities at a disadvantage, including the "last hired, first fired" policy and the reliance on property taxes for the funding of public schools; and (c) the results of the past 300 years of discrimination. For example, African Americans are disproportionately poor and unemployed and have little wealth compared with whites and hence are at a disadvantage due to the lower incomes they have, the less wealth they have, the poorer quality schools their children attend, the fewer family resources available to educate their children, and the poorer and more unsafe neighborhoods in which they live. Consequently, as Farley (1995) stated in *Majority–Minority Relations,* "American society has not yet attained the ideal of equal opportunity" (p. 443).

4. For two sources that seem to be saying that economic factors, rather than racial factors, are increasingly playing an influential role in the upward mobility of minorities, see Wilson's (1978) *The Declining Significance of Race,* where he stated that "the current problems of lower-class blacks are substantially related to fundamental structural changes in the economy" (pp. 21–22). See also Conley's (1999) *Being Black, Living in the Red.* Conley found that wealth, rather than skin color, had greater predictability of outcomes for minorities (p. 134) and that therefore there needs to be a corresponding "shift to a class-based affirmative action policy— that is, implementing educational, hiring, and contracting preferences that are based on *class* and not skin color" (p. 152, emphasis added).

5. Wilson (1987) stated that policies such as affirmative action are seen by many Americans as favoring minorities and discriminating against nonminorities, with the result that "the more the public programs are perceived by members of the wider society as benefiting only certain groups, the less support those programs receive"

(p. 118). Similar to Wilson's argument, Fishkin (1983) asserted that social policy needs to be based on the principle of equality of life chances where those who are truly disadvantaged, regardless of their race/ethnicity, would be helped to be upwardly mobile.

References

Conley, D. (1999). *Being black, living in the red: Race, wealth, and social policy in America.* Berkeley: University of California Press.

Farley, J. E. (1995). *Majority–minority relations* (3rd ed.). Englewood Cliffs, NJ: Prentice Hall.

Fine, G. A., & Shulman, D. (2003). Race and ethnicity: Should minorities be given preferential treatment in admission to higher education and hiring? In *Talking sociology* (5th ed., pp. 131–151). Boston: Allyn & Bacon.

Fishkin, J. S. (1983). *Justice, equal opportunity, and the family.* New Haven, CT: Yale University Press.

Gans, H. J. (1995). Joblessness and antipoverty policy in the twenty-first century. In *The war against the poor: The underclass and antipoverty policy* (pp. 133–147). New York: Basic Books.

Greenhouse, L. (2003, June 24). Affirmative action upheld, with limits. *Louisville Courier–Journal,* pp. A1, A4.

Kinsley, M. (2003, June 26). What diversity? Think fuzzy. *Louisville Courier–Journal,* p. A7.

Merton, R. K. (1967). Manifest and latent functions. In *On theoretical sociology: Five essays, old and new* (pp. 73–138). New York: Free Press.

Thomas, C. (2003, June 26). Diversity: Not a compelling interest. *Louisville Courier–Journal,* p. A7.

Wilson, W. J. (1978). *The declining significance of race: Blacks and changing American institutions.* Chicago: University of Chicago Press.

Wilson, W. J. (1987). *The truly disadvantaged: The inner city, the underclass, and public policy.* Chicago: University of Chicago Press.

Yetter, D. (2003, June 23). Many needy parents denied child-care help: State's long waiting list may push some on welfare. *Louisville Courier–Journal,* pp. A1, A4.

CHAPTER 6

Notes

1. See "State Welfare Cutbacks" (2002) and, in Kentucky, Yetter (2003). The latter article noted that 25 states "are not able to serve all families who apply"

(p. A4). Several single mothers who were interviewed felt that, without child care subsidies and with their low wages, "they will be forced to quit jobs" (p. A4). With low wages and no child care subsidies, they regret that they must return to welfare. To the degree that this is a pattern throughout our country, our society could be better served by providing more subsidies for child care so that these mothers can continue in their jobs and have the chance for some kind of upward mobility.

2. See Blumberg (1984) and Turner (2003), where Turner reformulates Blumberg's theoretical ideas into a number of interrelated theoretical propositions (p. 184).

3. See Clawson and Gerstel (2004, pp. 94–95). Also, in this same edition, see two other articles that discuss paid maternity leaves: Gornick and Meyers (2004, pp. 100–102) and Christopher (2004, p. 109).

4. Christopher (2004) went on to note, "Macroeconomic factors were far more important influences on the unemployment rate" (p. 111).

References

Armas, G. C. (2004, June 4). In most jobs, it pays to be a man, report says. *Louisville Courier-Journal,* p. A1.

Blumberg, R. L. (1984). A general theory of gender stratification. *Sociological Theory, 2,* 23–101.

Christopher, K. (2004). Family-friendly Europe. In D. S. Eitzen & C. S. Leedham (Eds.), *Solutions to social problems* (3rd ed., pp. 107–111). Boston: Allyn & Bacon.

Clawson, D., & Gerstel, N. (2004). Caring for our young: Child care in Europe and the United States. In D. S. Eitzen & C. S. Leedham (Eds.), *Solutions to social problems* (3rd ed., pp. 90–97). Boston: Allyn & Bacon.

Crone, J. A. (1998). Poverty. In C. L. Banston III (Ed.), *Encyclopedia of family life* (pp. 1060–1065). Pasadena, CA: Salem Press.

Eitzen, D. S., & Sage, G. H. (2003). *Sociology of North American sport* (7th ed.). Boston: McGraw-Hill.

Eitzen, D. S., & Zinn, M. B. (2003). *Social problems* (9th ed.). Boston: Allyn & Bacon.

Farley, J. E. (1995). *Majority-minority relations* (3rd ed.). Englewood Cliffs, NJ: Prentice Hall.

Gornick, J. C., & Meyers, M. K. (2004). Support for working families: What the United States can learn from Europe. In D. S. Eitzen & C. S. Leedham (Eds.), *Solutions to social problems* (3rd ed., pp. 98–106). Boston: Allyn & Bacon.

Kerbo, H. R. (2003). *Social stratification and inequality: Class conflict in historical, comparative, and global perspective* (5th ed.). Boston: McGraw-Hill.

State welfare cutbacks will slash child care, job training programs. (2002, September 11). *Madison Courier,* p. A2.

Turner, J. H. (2003). Feminist conflict theory. In *The structure of sociological theory* (7th ed., pp. 182–194). Belmont, CA: Wadsworth/Thomson Learning.

Yetter, D. (2003, June 23). Many needy parents denied child-care help. *Louisville Courier–Journal*, pp. A1, A4.

CHAPTER 7

Notes

1. In *Savage Inequalities,* Kozol (1991) observed how there was no staff to teach machine shop even though the school had a machine shop.

2. Before I went for my Ph.D. in sociology so that I would be able to teach at the college level, I taught for 3 years in a public high school. Even though I had bachelor's and master's degrees in world history as a major and sociology and political science as minors, I never taught a world history course, but I did teach an economics course even though I had taken only one college economics class before I taught the economics course. So, the subject area in which I had an interest and depth of knowledge of the subject matter, I never taught. But a subject area in which I had little interest and little knowledge of the subject matter, I was assigned to teach.

3. The National Education Association (NEA, n.d. a) asserted, "It is unacceptable for teachers to be assigned out-of-field. Such assignments are a disservice to students and teachers alike."

4. The NEA (n.d. a) asserted, "Teacher compensation is a significant deterrent to recruitment. Teachers are still paid less than professions that require comparable education and skills. Teachers still are not valued and respected to the extent of their actual contributions to society." Moreover, not only is recruiting teachers a challenge due to their low pay, but keeping them is also a problem. The NEA reported that "20 percent of all new hires leave the classroom within three years." In urban districts, "close to 50 percent of newcomers flee the profession during their first five years of teaching." Besides low compensation, those who leave say "they feel overwhelmed by the expectations and scope of the job. Many say they feel isolated and unsupported in their classrooms or that expectations are unclear."

5. Kozol (1991) noted how in schools in East St. Louis, Illinois, 280 teachers were laid off along with 166 cooks, 25 teacher aides, and 18 painters, electricians, engineers, and plumbers (p. 24). Loss of 280 teachers caused the size of the East St. Louis classes to get larger, which is the opposite of what needs to happen in a low-income, poverty-ridden city. He noted that the teachers' paychecks were arriving 2 weeks late and stated, "The city warns its teachers to expect a cut of half their pay until the fiscal crisis has been eased" (p. 24). Also, Fruchter (1998) noted, "We have sufficient evidence that reducing class size and providing intensive professional

development, particularly in the early grades, significantly increases student achievement" (p. 15).

6. The American Federation of Teachers (AFT, n.d. a) reported, "The Tennessee STAR [Student Teacher Achievement Ratio] study followed a group of students from kindergarten through third grade, randomly assigning these students to one of three types of classes: small (13–17 students); regular (22–25 students); and regular with an aide. With four years of data, researchers found that students in small classes significantly outperformed the other students in both math and reading, every year, at all grade levels, across all geographic areas."

7. The National Education Association (n.d. b) reported research on a longitudinal study done in Tennessee titled Student Teacher Achievement Ratio (STAR).

8. See NEA (n.d. b). The NEA also quoted the U.S. Department of Education: "A growing body of research demonstrates that students attending small classes in the early grades make more rapid educational progress than students in larger classes, and that these achievement gains persist well after students move on to larger classes in later grades."

9. The AFT (n.d. a) noted that with respect to the state of Tennessee's STAR research on the relationship between class size (the independent variable) and reading and math scores (the dependent variables), "lower class size makes a big difference for children, particularly poor children." The AFT stated that lowering class size in combination with "high academic standards and a challenging curriculum, safe and orderly classrooms, and qualified teachers [is] also necessary."

10. The NEA (n.d. c) stated, "Every child deserves the opportunity to learn in a classroom that is modern and equipped with the latest in educational technology. Too many students—one in three, according to the U.S. General Accounting Office—attend classes in schools that fail to meet these criteria." The NEA further noted that a study by the U.S. General Accounting Office some years ago estimated that it would take $112 billion to restore our nation's schools to "good overall condition," but given the intervening years and given the increasing enrollments and greater costs in new school construction, the NEA estimates that the cost now would be approximately $200 billion. Besides the disrepair and the overcrowding, the NEA noted that "forty-six percent of the public schools in America lack the electrical and communication wiring to support today's computer systems."

11. Kozol (1991) stated that when he traveled throughout the country during the years of 1988 and 1989 and visited many public schools, "Looking around some of these inner-city schools, where filth and disrepair were worse than anything I'd seen in 1964, I often wondered why we would agree to let our children go to school in places where no politician, school board president, or business CEO [chief executive officer] would dream of working" (p. 5). The football field had a couple of metal pipes but no crossbar for goalposts, and the school did not have a washing machine to wash the football uniforms, which were 9 years old (p. 25). Kozol noted how there was no heat in the weight room and no staff to teach machine shop even though the school had a machine shop and that the science labs "are 30 to 50 years outdated." The physics lab had no equipment, and the typewriters did not work (p. 30).

12. Kozol (1991) noted that the teachers in East St. Louis "are running out of chalk and paper" (p. 24).

13. Kozol (1991) discussed the lack of safety in the schools: "The doors [of the schools] were guarded. Police sometimes patrolled the halls. The windows of the schools were often covered with steel grates. Taxi drivers flatly refused to take me to some of these schools" (p. 5).

14. Hooper (2005) reported that the Indianapolis public schools are creating smaller high schools to have higher quality education and to keep more students in school. She reported that one third of the students drop out and that one third graduate without skills to go to college or to work (p. B5). Her article went on to report that the billionaire Bill Gates "believes so strongly that smaller is better that the foundation he and his wife support is sinking billions into helping districts make the change" (p. B5).

15. The NEA (n.d. d) took the stance that one way to make schools safer is to "provide resources for smaller classes and smaller schools." Also, Fruchter (1998) stated that we have "a growing mass of evidence that reducing scale by creating smaller schools raises student achievement, reduces dropping out, and increases graduation rates at the high school level" (p. 15).

16. Turner (2003, pp. 134–136) discussed Max Weber's proposition that as there are lower rates of mobility among subordinates of a social system, these subordinates will be more likely to withdraw their belief in the legitimacy of the current political authority and be more likely to pursue conflict with superordinates.

17. See Friedman (1982) for an early statement of the voucher system, where education would be taken out of the hands of government more and more and where individual citizens would choose where and what kind of education they wanted their children to have. In other words, Friedman wanted education to be more in the realm of the marketplace than under the authority of the government.

18. The AFT (n.d. c) noted, "Private and religious schools currently have almost complete autonomy with regard to who they teach." The public would want any private and religious schools to be accountable given that these schools would be using public money. Yet in research cited in this article, the AFT noted that these private and religious schools "would not be willing to participate in a voucher plan that requires them to meet the kind of accountability standards that the public desires." This report found that in the school districts studied, even if no conditions were made on these schools, the private and religious schools could accommodate at most only 3.5% of the public school enrollment. This would mean that 96.5% of the students would be left behind, still attending poor-quality public schools. Hence, we would be far from solving our public education problem.

19. A newspaper editorial ("After the Voucher Ruling," 2002) noted that in the situation in Cleveland, Ohio, "despite legislative intent, suburban public schools and well-off private ones are not participating" (p. A8). The editorial went on to assert that Kentucky and Indiana are finding more promising ways "than spending a lot of money to send relatively few students to religious schools that are publicly unaccountable and economically marginal" (p. A8). It made a final point that is

worth considering: "Further, there's a real threat of creating a vicious circle. Public schools could become even more overburdened and undersupported as they are left with even higher proportions of the special education or behaviorally troubled youngsters private schools won't accept or keep" (p. A8).

20. Gutmann (2000) stated, "If we [the citizens of this country] were committed to giving poor parents what most parents want for their children, we would not follow the voucher route; we would do whatever it takes to improve public schools" (p. 20).

21. Gutmann (2000) asserted, "A successful voucher movement in this country would therefore provide an enormous subsidy to affluent parents" (p. 20).

22. The AFT (n.d. b) reported on a national representative sample of more than 1,000 adults who were asked the question of whether we should allow students and parents to choose a private school to attend at public expense; the survey showed that 44% of the respondents favored this and 50% were against it. Moreover, when asked how the private schools should be accountable with this public tax money, 92% of the respondents said there should be no discrimination based on race, 88% said the private schools should meet state curriculum standards (as of that publication, "No state requires private schools to meet the same state curriculum standards as public schools"), 86% said only certified teachers should be employed (71% of all private school teachers are licensed vs. 97.4% of all public school teachers who are licensed), and 83% said the private schools should not discriminate on the basis of religious beliefs.

23. See NEA (n.d. e). Also, the AFT (n.d. b) noted that California's "proposal to give $4,000 vouchers for students to attend private or religious schools will be a $3 billion windfall to affluent parents whose children already attend private schools." Luis Huerta, coauthor of the report used by the AFT, stated, "It's essentially tax relief for the well off."

References

After the voucher ruling. (2002, June 29). *Louisville Courier–Journal*, p. A8.

American Federation of Teachers. (n.d. a). *Small class size: Education reform that works*. [Online.] Available: www.aft.org

American Federation of Teachers. (n.d. b). *Voucher home page reports*. [Online.] Available: www.aft.org

American Federation of Teachers. (n.d. c). *Vouchers and the accountability dilemma*. [Online.] Available: www.aft.org

Carroll, R. (2003, August 20). Most oppose vouchers, want teachers paid more. *Madison Courier*, p. A9.

Friedman, M. (1982). The role of government in education. In *Capitalism and freedom* (pp. 85–107). Chicago: University of Chicago Press.

Fruchter, N. (1998, Fall/Winter). American public education: Crisis and possibility. *New Labor Forum*, pp. 9–16.

Gutmann, A. (2000, Summer). What does "school choice" mean? *Dissent*, pp. 19–24.

Hooper, K. (2005, August 21). Less is more for students in Indy: Overhaul creates smaller schools. *Louisville Courier–Journal*, pp. B1, B5.

Kerbo, H. R. (2003). *Social stratification and inequality: Class conflict in historical, comparative, and global perspective* (5th ed.). Boston: McGraw–Hill.

Kozol, J. (1991). *Savage inequalities: Children in America's schools*. New York: Crown.

Merton, R. K. (1968). Social structure and anomie. In *Social Theory and Social Structure* (enlarged ed., pp. 185–214). New York: Free Press.

Mills, C. W. (1959). *The sociological imagination*. London: Oxford University Press.

National Education Association. (n.d. a). *Attracting and keeping quality teachers*. [Online.] Available: www.nea.org

National Education Association. (n.d. b). *Class size*. [Online.] Available: www.nea.org

National Education Association. (n.d. c). *School modernization*. [Online.] Available: www.nea.org

National Education Association. (n.d. d). *School safety*. [Online.] Available: www.nea.org

National Education Association. (n.d. e). *Vouchers*. [Online.] Available: www.nea.org

Rodriguez, N. C. (2003, February 5). School systems consider job, program cuts. *Louisville Courier–Journal*, p. A1.

Turner, J. H. (2003). *The structure of sociological theory* (7th ed.). Belmont, CA: Wadsworth/Thomson Learning.

CHAPTER 8

Notes

1. For a more extended discussion of slavery and its consequences, see Franklin and Moss's (1994) *From Slavery to Freedom*.

2. Hraba (1994) discussed various Indian nations and how each of them had its own distinct way of life, including different customs, religions, dress, and ways of economic survival (pp. 202–213).

3. See Hraba (1994, pp. 219–231). For a good discussion of the current situation of Native Americans, see Aguirre and Turner's (1991, pp. 104–133) *American Ethnicity*.

4. See Merton's (1968, pp. 185–214) article in *Social Theory and Social Structure*. See also Cloward and Ohlin's (1960) *Delinquency and Opportunity*.

5. Eitzen and Leedham (2004) asserted that one of the major causes of crimes, especially violent crimes such as robbery, assault, murder, and rape, is the greater proportion of Americans living in poverty compared with rates of poverty and crime

in other industrialized countries. They also suggested that a second cause of more violent crime can be seen when a country has a weaker safety net for the poor such as we do compared with other industrialized countries (p. 190).

6. Eitzen and Leedham (2004) asserted that one of the four factors that cause crime, especially violent crime, is a large gap between the rich and the poor. They noted, "The United States has the greatest inequality gap among industrialized nations" (p. 190).

References

Aguirre, A., Jr., & Turner, J. H. (2001). *American ethnicity: The dynamics and consequences of discrimination* (3rd ed.). Boston: McGraw–Hill.

Booth, M. (2004). Deaths reflect gun use in U.S. In D. S. Eitzen & C. Leedham (Eds.), *Solutions to social problems: Lessons from other societies* (3rd ed., pp. 197–199). Boston: Allyn & Bacon.

Cloward, R. A., & Ohlin, L. E. (1960). *Delinquency and opportunity: A theory of delinquent gangs.* New York: Free Press.

Durkheim, E. (1966). What is a social fact? In *The rules of sociological method* (pp. 1–13). New York: Free Press.

Eitzen, D. S., & Leedham, C. S. (2004). Crime and crime control. In *Solutions to social problems: Lessons from other societies* (3rd ed., pp. 190–208). Boston: Allyn & Bacon.

Franklin, J. H., & Moss, A. A., Jr. (1994). *From slavery to freedom: A history of African Americans* (7th ed.). New York: McGraw–Hill.

Hraba, J. (1994). *American ethnicity* (2nd ed.). Itasca, IL: F. E. Peacock.

Marx, K. (1964). Estranged labor. In *The economic and philosophic manuscripts of 1844* (pp. 106–119). New York: International Publishers.

Marx, K. (1967). *Capital.* New York: International Publishers.

Marx, K., & Engels, F. (1992). *The Communist manifesto.* New York: Bantam.

Merton, R. K. (1968). Social structure and anomie. In *Social theory and social structure* (enlarged ed., pp. 185–214). New York: Free Press.

Paternoster, R. (1989). Absolute and restrictive deterrence in a panel of youth: Explaining the onset, persistence/desistance, and frequency of delinquent offending. *Social Problems, 36,* 289–309.

Pear, R. (1999, October 4). 44.3 million have no health insurance. *Louisville Courier–Journal,* p. A1.

Population Reference Bureau. (2002). *2002 world population data sheet.* Washington, DC: Author.

Ritzer, G. (2005). *Enchanting a disenchanted world: Revolutionizing the means of consumption* (2nd ed.). Thousand Oaks, CA: Pine Forge.

University of Texas. (n.d.). *Unemployment rate, Austin–Round Rock MSA, Texas, and U.S., 1999–2003.* [Online.] Available: www.utexas.edu/depts/bbr/austindex/snapshot/unemployment/unemprate.pdf

CHAPTER 9

Notes

1. Nadelmann (1988) went on to say, "All of the health costs of marijuana, cocaine, and heroin combined amount to only a small fraction of those caused by tobacco and alcohol" (p. 24).

2. Brownstein (1992) noted, "Efforts to reduce the supply of imported drugs through interdiction have resulted in an expansion of the domestic supply of marijuana, more violence, and possibly even assistance to major drug dealers through elimination of weaker competitors" (p. 220).

3. Nadelmann (1988) noted, "Many illicit drug users commit crimes such as robbery and burglary, as well as drug dealing, prostitution, and numbers running, to earn enough money to purchase the relatively high-priced illicit drugs. . . . If the drugs to which they are addicted were significantly cheaper—which would be the case if they were legalized—the number of crimes committed by drug addicts to pay for their habits would, in all likelihood, decline dramatically. Even if a legal drug policy included the imposition of relatively high consumption taxes in order to discourage consumption, drug prices would probably still be lower than they are today" (p. 17).

4. Nadelmann stated, "Illegal markets tend to breed violence. . . . During Prohibition, violent struggles between bootlegging gangs and hijackings of booze-laden trucks and sea vessels were frequent and notorious occurrences. . . . Most law enforcement officials agree that the dramatic increases in urban murder rates during the past few years [this article was published in 1988] can be explained almost entirely by the rise in drug dealer killings" (p. 18).

5. Nadelmann (1988) noted, "Repeal of Prohibition came to be seen not as a capitulation of Al Capone and his ilk, but as a means of both putting the bootleggers out of business and eliminating most of the costs associated with the Prohibition laws" (p. 13).

6. Szasz (1972) argued, "Like most rights, the right of self-medication should apply only to adults" (p. 77).

7. Nadelmann (1988) noted, "If the marijuana, cocaine, and heroin markets were legal, state and federal governments would collect billions of dollars annually in tax revenues" (p. 16).

8. Becker (2001) also suggested this approach because "legalizing drugs is a venture into the unknown" (p. 32).

9. Brown (1975) noted that the National Commission on Marijuana and Drug Abuse, created by Congress in 1970, recommended that the possession of marijuana "for personal use no longer be a federal or state offense" (p. 114).

10. Eitzen and Zinn (2003) cited that it costs $80,000 to build for each new cell and costs approximately $25,000 per year to house a prisoner (p. 390).

11. Nadelmann (1988) agreed. He stated, "Legalization is repeatedly and vociferously dismissed, without any attempt to evaluate it openly and objectively. The past twenty years have demonstrated that a drug policy shaped by exaggerated rhetoric designed to arouse fear has only led to our current disaster. Unless we are willing to honestly evaluate our options, including various legalization strategies, we will run a still greater risk: We may never find the best solution for our drug problems" (p. 31).

12. See Cart (1999). New Mexico Governor Gary Johnson advocated legalizing all drugs and suggested that this issue needs to be discussed, but he heard a lot of negative criticism just for bringing up the issue.

References

Becker, G. (2001, September 17). It's time to give up the war on drugs. *Business Week,* p. 32.

Brown, B. S. (1975). Drugs and public health: Issues and answers. *Annals of the American Academy of Political and Social Sciences, 417,* 110–119.

Brownstein, H. H. (1992). Making peace in the war on drugs. *Humanity and Society, 16,* 217–235.

Cart, J. (1999, November 12). Governor's support for drugs sparks New Mexico firestorm. *Louisville Courier–Journal,* p. A4.

Currie, E. (1993, Winter). Toward a policy on drugs: Decriminalization? Legalization? *Dissent,* pp. 65–71.

Eitzen, D. S., & Zinn, M. B. (2003). *Social problems* (9th ed.). Boston: Allyn & Bacon.

Fine, G. A., & Shulman, D. (2003). *Talking sociology* (5th ed.). Boston: Allyn & Bacon.

Myers, L. (2001). Cultural divide over crime and punishment. In D. S. Eitzen & C. S. Leedham (Eds.), *Solutions to social problems: Lessons from other societies* (2nd ed., pp. 233–239). Boston: Allyn & Bacon.

Nadelmann, E. A. (1988, Summer). The case for legalization. *The Public Interest,* pp. 3–31.

Szasz, T. S. (1972, April). The ethics of addiction. *Harper's,* pp. 74–79.

CHAPTER 10

Notes

1. This was reported on the *NewsHour with Jim Lehrer,* January 9, 2006. One interviewee reported that 46 million Americans are currently uninsured. It was also

reported that 69% of American corporations offered health coverage in 2000, whereas only 60% of American corporations offer coverage today. Because health insurance costs so much, and because the cost is going up at 8% or more each year (as reported on the *NewsHour*), corporations are finding it harder and harder to offer coverage. To cut their costs so as to remain competitive or become more competitive in their respective markets, more and more corporations are reducing or eliminating health insurance coverage for their employees. This trend will put more pressure on the current president and future presidents, as well as the current Congress and future Congresses, to do something in the area of health care for Americans.

2. Unger (2006) noted, "Many people cannot afford to see a doctor and live with the threat of financial ruin if they get sick" (p. A1).

3. See Carroll (2003, p. A1). By the way, the new prescription drug plan took effect January 1, 2006, not without some problems. Pear (2006b) noted that people who had signed up were not on the government's list of subscribers, that insurers did not have a way of identifying poorer people who were entitled to extra help, and that pharmacists were on the phone for hours trying to reach insurers to work out the various problems. The new program, Pear noted, appeared to be "too complicated for many people to understand" (p. A12). He stated that this new prescription drug plan "is the most significant expansion of Medicare since creation of the program in 1965" (p. A12). Also, Howington (2006) reported that in 70% of the cases in Kentucky, there were problems in getting prescriptions (p. A1). Retired people were not on the new lists even though they had signed up to be on the lists, and poor people who are on Medicaid were supposed to get their prescription drugs through Medicare, but "Medicare plans wouldn't pay" (p. A1). A major problem, the article noted, was that there was not enough time to sign up 42 million Americans between the sign-up time of November 15, 2005, and the time the new prescription drug plan kicked in on January 1, 2006.

4. Jewell (2003) reported that Republican congresspersons, such as Dan Burton of Indiana and Jo Ann Emerson of Missouri, desired the importation of prescription drugs as a way to decrease costs of prescription drugs for the elderly. This article reported that the Eli Lilly drug company, with headquarters in Indianapolis, Indiana, got 100 of its employees to go to a forum promoting the importation of drugs to speak against this importation. Also, Frommer (2003) reported that the drug industry spent $29 million in lobbying—"more than any other industry" (p. A10). Besides the lobby that spent $8.5 million to work against allowing Americans to buy prescription drugs from Canada, individual drug companies also spent their own money to stop the importation of prescription drugs at lower prices; for example, Eli Lilly spent $2.9 million, Bristol–Myers Squibb spent $2.6 million, Johnson & Johnson spent $2.2 million, and Pfizer spent $1.8 million. Ira Loss, a senior health care analyst for the Washington Analysis Corporation, stated, "They [the drug companies] make a lot of money, and they are trying to protect their interests" (p. A10). Frommer reported that the drug industry "enjoyed about $150 billion in U.S. sales last year [the article was published in 2003]" (p. A10).

5. Graig (1999) noted, "Many U.S. companies complained that high health care costs put them at a competitive disadvantage with regard to foreign competitors in countries with nationalized health care systems" (p. 18). Graig reported further that although few employers viewed health care costs as a problem in 1980, two thirds of the executives in a 1990 survey viewed "health benefits costs to be the leading issue companies faced at that time" (p. 19).

6. This was reported on the *NewsHour with Jim Lehrer,* January 12, 2006.

7. Kerbo (2006) noted that good health is unequally distributed in that people with lower incomes will often face the following conditions: (a) poor nutrition, (b) less sanitary living conditions, (c) less knowledge about how to have better health, and (d) more unhealthy work environments such as working with dangerous machines and chemicals (p. 38).

8. Also, Globerman (1990) noted that there are physicians who are "committed to the well-being of the collective, who support government-run universal health insurance" (p. 12), but that there are also physicians who "oppose government intervention and support free-enterprise medicine" (p. 12). For the physicians who want more free enterprise medicine, the reasons why they want this are "professional ideology, free enterprise ideology, and economic self-interest" (p. 13). So, even though Canada has a universal health care system, there are still doctors who want to abolish this system. A number of these doctors want to bill the patients over and above what the government will reimburse them for services rendered, for example, approximately 12% of doctors' "extra-bill" patients (p. 15).

9. Pear and Bogdanich (2003) reported, "Drug companies say they support covering prescription drugs under Medicare. But in the last few years, they have invested several hundred million dollars in campaign contributions, lobbying, and advertising to head off price controls" (p. A7).

10. In a classic work in sociology, *The Social Construction of Reality,* Berger and Luckmann (1966) argued that we, as humans, create our own social reality of norms, values, beliefs, and institutions. These norms, values, beliefs, and institutions may be different in different cultures and at different times in history, but they are created by humans over time. Sometimes people forget this and believe that these things have always been and will always be. Berger and Luckmann reminded us how much we socially construct how we live.

11. Also, Daniels and colleagues (1996) noted, "In poll after poll, over many years, an overwhelming majority of Americans say they believe no one should be denied needed medical services because of an inability to pay" (p. 16).

12. See Walker (1969). There was favorable interest in creating a state or national health care system in the United States during the early 1900s, but Walker noted that "interest never penetrated to the level of the workingman" (p. 304).

13. Daniels and colleagues (1996) presented data showing that "the lowest fifth of the nation has no money left over for medical bills, health insurance premiums, movie going, or any other expenses" (p. 50). So, the lowest 20% of Americans do not make enough money to have health insurance. Moreover, Daniels and colleagues pointed out that the next 20% of Americans do not have much left over

after paying their expenses. So, in reality, approximately 40% of Americans either cannot afford health care coverage or have a difficult time in making enough money to pay for health insurance.

14. Graig (1999) discussed the first national health care plan begun under the German leader Otto von Bismarck in 1871. Graig noted that in Bismarck's health care system, "all members of society should have access to health care regardless of ability to pay, and . . . the costs of health care would be spread across the population" (p. 47). Bismarck was responding to the rise of socialism and wanted to retain the loyalty of the workers by implementing such a health care plan. People paid into this plan via their employers and paid more or less depending on their incomes (p. 47).

15. Daniels and colleagues (1996) asserted that "any health reform must retain a strong emphasis on public health and prevention" (p. 54).

16. Others have come up with similar criteria. For example, Daniels and colleagues (1996) discussed 10 criteria, some of them similar to my criteria. For example, they stated that a new health care system needs to be accessible to all (my Criterion 2), that it needs to be comprehensive in benefits and be uniform in benefits (similar to my Criteria 3 and 4 where all Americans will receive good-quality care and receive it in a timely way), and that it needs to be accountable to the public (similar to my Criteria 8, 9, and 10 where health professionals, drug companies, hospitals, and insurance companies are paid reasonably or are compensated reasonably) (pp. 32–34).

17. See Health Canada (n.d. a), where the Canadians solve the problem of a waiting period by having the province the person is moving from continue to pick up the person's health care coverage for a 3-month waiting period when the person moves to another province and applies for health care coverage.

18. Yetter (2005) noted, "Children are doing without dental care because their parents don't have a car or other means to take them out of the county to a dentist" (p. A2).

19. We are already beginning to see preliminary attempts at providing more nutritious food at public schools. See, for example, Schneider's (2006) newspaper article.

20. This was reported on the *NewsHour with Jim Lehrer,* January 12, 2006.

21. See "Medicare Q & A" (2003, p. A7). Medicare is a federal program, whereas Medicaid is a joint federal and state program where the "programs vary from state to state." Also, Sherman (2003) noted that whereas the federal government pays for all of the Medicare bill, the "states pay about a third of the cost of Medicaid" (p. A5). From the state governors' perspectives, because most of them are facing large deficits in their state budgets, they are hoping that the federal government can take over the cost of prescription drugs given that this would help to cut the states' Medicaid costs, potentially by $7 billion.

22. For a succinct discussion of the libertarian view, see Fine and Shulman's (2003, pp. 6–9) *Talking Sociology.*

23. See Archer (1998, p. 77), who suggested changes that could be made in the current system even if we do not go to a national health care system that would

greatly improve managed health care: "(1) Pay managed care plans in a manner that creates a financial incentive to recruit patients in poor health and provide them with quality care, and (2) mandate that plans disclose information that can permit evaluation of plan treatment practices" (p. 78). Under this situation, the government would need to pay HMOs enough to give them an incentive to include seriously ill people in their plans. Also, the government would need to stipulate that all HMOs disclose what services they provide so that people can decide which HMO they want to join. Currently, people do not have this information with which to decide what is in their best interest. Also, Newhouse (1996) noted that one of the potential problems managed care raises is that "above average spenders, for example those with chronic illness, may be shunned by every plan" (p. 1719).

24. See Graig (1999). German public opinion "was enraged by the wide disparity in income levels between physicians and other workers" (p. 62). Doctors during the 1970s made five to six times more than the average worker.

25. Marmor (1996) outlined how there were four times in American history when there were attempts "to enact national health insurance" (p. 672). He went on to state, "Now, as then, entrenched interests tried to block national health insurance by skillfully manipulating our deepest fears to protect what they regarded as their interests" (p. 672).

26. See Pear (1999, p. A1). The 46 million figure was reported on the *NewsHour with Jim Lehrer,* January 9, 2006.

27. See Wheeler (2003, p. A4), Dalrymple (2003, p. A10), and Yetter (2003a, pp. A1, A6; 2003b, pp. A1, A6). See also Stolberg (2006, p. A1). By narrow margins in both the Senate and the House, a bill to cut spending on Medicare, Medicaid, and student loans passed. As a consequence, more elderly and poor will have problems receiving health care.

References

Archer, D. (1998, Summer). From a Medicare rights advocate: Problems and solutions in Medicare managed care. *Generations, 22*(2), 77–78.

Beauchamp, D. E. (1996). *Health care reform and the battle for the body politic.* Philadelphia: Temple University Press.

Berger, P. L., & Luckmann, T. (1966). *The social construction of reality: A treatise in the sociology of knowledge.* New York: Anchor Books.

Budrys, G. (2003). *Unequal health: How inequality contributes to health or illness.* Lanham, MD: Rowman & Littlefield.

Carroll, J. R. (2003, September 14). Canada's lure: Low-cost prescription drugs. *Louisville Courier–Journal,* pp. A1, A7.

Dalrymple, M. (2003, September 22). States trim Medicaid to balance budgets. *Madison Courier,* p. A10.

Daniels, N., Light, D. W., & Caplan, R. L. (1996). *Benchmarks of fairness for health care reform.* New York: Oxford University Press.

Durbin, D. A. (2005, November 22). UAW calls GM moves devastating. *Madison Courier,* pp. A1, A10.

Feldman, R. (2006, February 21). Ailing health system needs more than a few Band-Aids. *Indianapolis Star,* p. A6.

Fine, G. A., & Shulman, D. (2003). *Talking sociology* (5th ed.). Boston: Allyn & Bacon.

Frommer, F. J. (2003, October 13). Drug lobby spends heavily against drug importation bill. *Madison Courier,* p. A10.

Globerman, J. (1990). Free enterprise, professional ideology, and self-interest: An analysis of resistance by Canadian physicians to universal health insurance. *Journal of Health and Social Behavior, 31,* 11–27.

Graig, L. A. (1999). *Health of nations* (3rd ed.). Washington, DC: Congressional Quarterly Press.

Health Canada. (n.d. a). *Canada Health Act: Overview.* [Online.] Available: www.hc-sc.gc.ca/hcs-sss/medi-assur/overview-apercu/index_e.html

Health Canada. (n.d. b). *Health care system.* [Online.] Available: www.hc-sc.gc.ca/hcs-sss/index_e.html

Health savings accounts: Approach has pros and cons. (2006, February 1). *Louisville Courier–Journal,* p. A7.

Howington, P. (2006, February 1). Health savings accounts: HSAs more popular, draw criticism. *Louisville Courier–Journal,* pp. A1, A7.

Jewell, M. (2003, September 17). Lilly loudly fights drug importation plan. *Madison Courier,* p. A2.

Kerbo, H. R. (2006). *Social stratification and inequality: Class conflict in historical, comparative, and global perspective* (6th ed.). Boston: McGraw–Hill.

Lester, W. (2003, October 20). Poll finds support for changing health system, prescription laws. *Louisville Courier–Journal,* p. A5.

Marmor, T. R. (1996). The politics of universal health insurance: Lessons from the past? *Journal of Interdisciplinary History, 26,* 671–679.

Marx, K., & Engels, F. (1992). *The Communist manifesto.* New York: Bantam.

Maynard, M. (2006, January 24). Ford eliminating up to 30,000 jobs and 14 factories. *The New York Times,* pp. A1, C6.

Medicare Q & A. (2003, June 29). *Louisville Courier–Journal,* p. A7.

Newhouse, J. P. (1996). Health reform in the United States. *The Economic Journal,* pp. 1713–1724.

Pear, R. (1999, October 4). 44.3 million have no health insurance. *Louisville Courier–Journal,* p. A1.

Pear, R. (2006a, January 30). Budget will hurt poor's health care, report says: Medicaid, Medicare would face big cuts. *Louisville Courier–Journal,* p. A3.

Pear, R. (2006b, January 16). President tells insurers to aid new drug plan. *The New York Times,* pp. A1, A12.

Pear, R., & Bogdanich, W. (2003, September 5). Problems with Medicare drug bill. *Louisville Courier–Journal,* p. A7.

Schneider, M. B. (2006, February 21). Healthy choices at school. *Indianapolis Star,* pp. A1, A4.

Sherman, M. (2003, August 5). Governors urge federal deal to ease drug costs for states. *Louisville Courier–Journal*, p. A5.

Stolberg, S. G. (2006, February 3). House approves budget cutbacks of $39.5 billion. *The New York Times*, p. A1.

Unger, L. (2006, January 30). Support swells for universal health care. *Louisville Courier–Journal*, pp. A1, A10.

Walker, F. A. (1969). Compulsory health insurance: "The next great step in social legislation." *Journal of American History, 56*, 290–304.

Weber, M. (1968). The distribution of power within the political community: Class, status, party. In *Economy and society: An outline of interpretive sociology* (pp. 926–940). New York: Bedminster.

Weigel, J. G. (2006, January 30). Medicare's new prescription drug plan: VA plan already set up. *Louisville Courier–Journal*, p. A9.

Wheeler, L. (2003, September 23). States tightening spending on Medicaid even further. *Louisville Courier–Journal*, p. A4.

Yetter, D. (2003a, August 31). Burden of care overwhelms unprepared families. *Louisville Courier–Journal*, pp. A1, A6.

Yetter, D. (2003b, September 18). Panel urges halt to Medicaid cuts. *Louisville Courier–Journal*, pp. A1, A6.

Yetter, D. (2005, December 29). Poor youths lack dental care. *Louisville Courier–Journal*, pp. A1–A2.

CHAPTER 11

Notes

1. Kerbo (2006) noted, for example, that the "gap between the average worker's pay and that of top corporate executives has shown a staggering increase from 40 to 1 in 1990 to 419 to 1 in 1998" (p. 23). Also, in the fifth edition of Kerbo's (2003) book, see Tables 2-4 and 2-5, where worker pay ranks 9th among the 13 industrialized nations, whereas chief executive officer salaries are the highest among these industrialized nations (p. 31).

2. Groppe (n.d.) stated that Indiana is one of seven states "that tax the incomes of three- or four-person families earning less than three-quarters of the poverty line."

3. Kerbo (2006) noted that of 12 industrialized countries, the United States does the least to decrease poverty within its borders (p. 251).

4. See Merton (1968, pp. 185–214) and Cloward and Ohlin (1960). Both works made connections among the social structure, the opportunities available, and the crime that can result.

5. One of the theoretical propositions that Turner (2003) constructed from the works of Max Weber in constructing a theory of conflict deals with the notion that

as there are lower rates of social mobility, "subordinates are more likely to with-draw legitimacy from political authority" (p. 135). As people find less legitimacy in the existing social structure, they are more likely to look for other means to survive and get ahead in that social structure.

6. See LeVay's (1991) journal article. Also, for one of the first extensive studies on the sexual behavior of humans, see Kinsey, Pomeroy, and Martin's (1948) *Sexual Behavior in the Human Male* and Kinsey, Pomeroy, Martin, and Gebhard's (1953) *Sexual Behavior in the Human Female*. For a more recent study on the biology of homosexuality, see Hamer and Copeland's (1994) *The Science of Desire*. In the study on the sexual behavior of the human male, Kinsey and colleagues (1948) made the following pertinent point: "Males do not represent two discrete populations, heterosexual and homosexual. The world is not to be divided into sheep and goats. Not all things are black nor all things white. It is a fundamental of taxonomy that nature rarely deals with discrete categories. Only the human mind invents categories and tries to force facts into separated pigeon-holes. The living world is a continuum in each and every one of its aspects. The sooner we learn this concerning human sexual behavior, the sooner we shall reach a sound understanding of the realities of sex" (p. 639).

7. See the two studies by Kinsey and colleagues (1948, 1953), where in actuality individual Americans are various degrees of being heterosexual and homosexual.

8. For more recent research on sexuality and how complex it can be (e.g., people can be attracted to the same sex, people can engage in sexual behavior with the same sex, people can identify with a certain sexual orientation or some combination of these dimensions), see Michael, Gagnon, Laumann, and Kolata's (1994) *Sex in America*.

References

Cloward, R. A., & Ohlin, L. E. (1960). *Delinquency and opportunity: A theory of delinquent gangs*. New York: Free Press.

Cooley, C. H. (1964). *Human nature and the social order*. New York: Schocken Books.

Groppe, M. (n.d.). Many poor Hoosiers struggling, groups say. *Indystar.com*. [Online.] Available: www.indystar.com/apps/pbcs.dll/article?aid=/20060224/news02/602240431

Hamer, D., & Copeland, P. (1994). *The science of desire: The search for the gay gene and the biology of behavior*. New York: Simon & Schuster.

Kerbo, H. R. (2003). *Social stratification and inequality: Class conflict in historical, comparative, and global perspective* (5th ed.). Boston: McGraw–Hill.

Kerbo, H. R. (2006). *Social stratification and inequality: Class conflict in historical, comparative, and global perspective* (6th ed.). Boston: McGraw–Hill.

Kinsey, A. C., Pomeroy, W. B., & Martin, C. E. (1948). *Sexual behavior in the human male*. Philadelphia: W. B. Saunders.

Kinsey, A. C., Pomeroy, W. B., Martin, C. E., & Gebhard, P. H. (1953). *Sexual behavior in the human female*. Philadelphia: W. B. Saunders.

Kozol, J. (1991). *Savage inequalities: Children in America's schools*. New York: Crown.

LeVay, S. (1991). A difference in hypothalamic structure between heterosexual and homosexual men. *Science, 253*, 1034–1037.

Liebow, E. (1967). *Tally's corner: A study of Negro streetcorner men*. Boston: Little, Brown.

Macionis, J. J. (2005). *Sociology* (10th ed.). Upper Saddle River, NJ: Pearson/Prentice Hall.

Merton, R. K. (1968). Social structure and anomie. In *Social theory and social structure* (enlarged ed., pp. 185–214). New York: Free Press.

Michael, R. T., Gagnon, J. H., Laumann, E. O., & Kolata, G. (1994). *Sex in America: A definitive survey*. Boston: Little, Brown.

Turner, J. H. (2003). Early conflict theory. In *The structure of sociological theory* (7th ed., pp. 131–161). Belmont, CA: Wadsworth/Thomson Learning.

CHAPTER 12

Notes

1. For interesting data to study on this matter and for a comparative look both among and between countries and a look over time, see the *1990 World Population Data Sheet* (Population Reference Bureau, 1990), which reported that the world's doubling time in 1990 was every 39 years. The *1997 World Population Data Sheet* (Population Reference Bureau, 1997) reported that the doubling time was 47 years. The *2003 World Population Data Sheet* (Population Reference Bureau, 2003) showed a 1.3% increase in population per year, meaning that the world's population is doubling every 54 years.

2. As we have learned since the onset of the war with Iraq, when Saddam Hussein was in power, he favored the Sunni ethnic group in Iraq while discriminating against the Shiites and Kurds with respect to the distribution of resources in that country. The unequal distribution of resources has been a key reason for the conflict in Northern Ireland between the Catholics and the Protestants. The same is true after the breakup of Yugoslavia, where the Serbian ethnic group came to have more power and more resources than did other ethnic groups (Gelles & Levine, 1999, pp. 313–315). During the past few years, we have seen people from the Darfur region of Sudan being beaten, killed, and run out of their homes and have seen their homes and entire villages being set afire by rebel forces and, at times, with the help of the Sudanese government. Brinkley and Polgreen (2006) reported, "More than 200,000 people have died and two million more have been driven from their homes since the conflict began in February, 2003" (p. A3).

3. For example, in western, central, and eastern Africa, families are having approximately six children per family, yet the average income of these families is less than $2,000 per year and some countries have an average income of less than $1,000 per year. See Population Reference Bureau (2003).

4. Forrester (1975) noted that a lot of countries and families would highly resist international controls on them in terms of family planning. Hence, providing access, education, and transportation would help not only to decrease the birth rate but also to give people freedom.

5. For the current George W. Bush administration, see Associated Press (2002, p. A4). For the past George Bush administration of the late 1980s and early 1990s, see Gore (2000, pp. 315–316).

6. Livernash and Rodenburg (1998) discussed the problems between developed and developing countries and possible solutions to world problems, where the developing nations have emphasized the need to forgive debt but the developed countries have emphasized family planning policies to decrease the population (pp. 9–10).

References

Associated Press. (2002, July 23). U.S. will withhold $34 million from U.N. birth-control program. *Louisville Courier–Journal*, p. A4.

Brinkley, J., & Polgreen, L. (2006, May 2). U.S. diplomat heads to Nigeria to try to unsnarl Darfur talks. *The New York Times*, p. A3.

Eitzen, D. S., & Zinn, M. B. (2000). *Social problems* (8th ed.). Boston: Allyn & Bacon.

Epstein, J. H. (1998, October). Declining growth in population raises hopes. *The Futurist*, p. 8.

Forrester, J. W. (1975, October). The road to world harmony. *The Futurist*, pp. 231–234.

Gelles, R. J., & Levine, A. (1999). *Sociology: An introduction* (6th ed.). Boston: McGraw–Hill.

Gore, A. (2000). *Earth in the balance: Ecology and the human spirit*. Boston: Houghton Mifflin.

Kerbo, H. R. (2006). *World poverty: Global inequality and the modern world system*. Boston: McGraw–Hill.

Livernash, R., & Rodenburg, E. (1998, March). Population change, resources, and the environment. *Population Bulletin*, pp. 2–40.

Population Reference Bureau. (1990). *1990 world population data sheet*. Washington, DC: Author.

Population Reference Bureau. (1997). *1997 world population data sheet*. Washington, DC: Author.

Population Reference Bureau. (2003). *2003 world population data sheet*. Washington, DC: Author.

Rosenfield, A. G. (2000). After Cairo: Women's reproductive and sexual health, rights, and empowerment. *American Journal of Public Health, 90,* 1838–1840.

Stoltenberg, T. (1989). Devising world policy options for sustainable growth. *Development, 1,* 19–25.

CHAPTER 13

Notes

1. See Gore (2000, p. 30). For the human authorship of much of our current environmental problems, see also Lori Hunter's (2001) *The Environmental Implications of Population Dynamics.*

2. For an excellent article on the pollution and waste problems of Mexico City, Mexico, see Harris and Puente (1990).

3. Gore (2000) stated that during the 20th century, "the average global surface temperature climbed one degree Fahrenheit and sea levels rose four to ten inches" (p. xiv). He also noted that data collected indicate that "the north polar cap has thinned 2 percent in just the last decade" (p. 23).

4. Romm (1991) reported that a National Academy of Sciences study, "which predicted a 1–5 degree rise, refused to rule out the possibility of a runaway greenhouse effect, with 'altered weather patterns' and 'a sea level several meters higher than it is today.'" (p. 35).

5. Romm (1991) stated that in our post-cold war world, "it is clear that America's current practice of focusing on short-term military threats will be singularly inadequate for dealing with broader, longer-term threats to the environment and the protection of vital resources" (p. 31).

6. Boeker and Van Grondelle (2000) discussed what will probably happen with the Kyoto Protocol and how, instead of counting national emissions by each country, the emissions should be connected to where the profit is; otherwise, developed countries could put more of their companies in developing countries, and the emissions would be counted as being produced by those countries rather than by the developed countries (p. 84).

7. Murshed (1993) asserted that the process of international agreement making will be crucial if we are to address and solve our environmental problems (p. 41).

8. The final chapter of Gore's (2000) book, titled "A Global Marshall Plan" (pp. 295–360), has an extensive set of suggestions on how we can solve the environmental problem. Anyone who seriously wishes to solve the environmental problem needs to read and consider what Gore has proposed.

9. Blumenthal (n.d.) reported, "The U.S. Army Corps of Engineers proposed to study how New Orleans could be protected from a catastrophic hurricane, but the

Bush administration ordered that the research not be undertaken." Blumenthal went on to say, "By 2003 the federal funding for the flood control project essentially dried up as it was drained into the Iraq war." He also noted that in 2004, "the Bush administration cut funding requested by the New Orleans district of U.S. Army Corps of Engineers for holding back the waters of Lake Pontchartrain by more than 80 percent." So, the chaos for not planning ahead was overwhelming in this instance. Hence, there is a critical need for planning and research and for setting aside funding to address potential catastrophes.

10. See Gore (2000, p. xvii). Gore took this position in opposition to what then Texas Governor George W. Bush would eventually do as U.S. president.

11. See Carroll (2002a, p. A10). Thelma Wiggins, spokeswoman for the Nuclear Energy Institute, was quoted as follows: "'The nuclear industry has an impeccable safety record,' she said. 'We have been transporting fuel for more than 35 years. There's been more than 3,000 shipments covering 1.7 million miles with no injuries, no fatalities, and no injury to the environment.'"

12. For an excellent discussion on vested interests and the influence of economic interests over environmental interests, see Van Der Straaten and Ugelow (1993).

References

Associated Press. (2002, July 24). Bush signs bill for Yucca Mountain dump. *Louisville Courier–Journal,* p. A10.

Blumenthal, S. (n.d.). No one can say they didn't see it coming. *Spiegel Online.* [Online English site.] Available: http://service.spiegel.de/cache/international/0,1518,372455,00.html

Boeker, E., & Van Grondelle, R. (2000). The environment as a human right. *International Journal of Human Rights, 4,* 74–93.

Carroll, J. R. (2002a, July 7). Large amount of nuclear waste could pass through area, analysis shows. *Louisville Courier–Journal,* pp. A1, A10.

Carroll, J. R. (2002b, July 10). Senate backs nuclear-waste storage site. *Louisville Courier–Journal,* pp. A1, A5.

Durbin, D. A. (2005, November 22). UAW calls GM moves devastating. *Madison Courier,* pp. A1, A10.

Ehrlich, P. R., Daily, G. C., Daily, S. C., Myers, N., & Salzman, J. (1997, December). No middle way on the environment. *Atlantic Monthly,* pp. 98–104.

Eitzen, D. S., & Leedham, C. S. (2001). *Solutions to social problems: Lessons from other societies* (2nd ed.). Boston: Allyn & Bacon.

Eitzen, D. S., & Zinn, M. B. (2000). *Social problems* (8th ed.). Boston: Allyn & Bacon.

Friedman, T. (2006, January 28). What Bush's speech should say. *Louisville Courier–Journal,* p. A9.

Gore, A. (2000). *Earth in the balance: Ecology and the human spirit.* Boston: Houghton Mifflin.

Harris, N., & Puente, S. (1990). Environmental issues in the cities of the developing world: The case of Mexico City. *Journal of International Development, 2,* 500–532.

Hunter, L. (2001). *The environmental implications of population dynamics.* Santa Monica, CA: RAND.

Intergovernmental Panel on Climate Change. (2001). *Climate change 2001: Impacts, adaptation, and vulnerability.* New York: United Nations Environmental Program and World Meteorological Organization.

Jan, G. (1995). Environmental protection in China. In O. P. Dwivedi & D. K. Vajpeyi (Eds.), *Environmental policies in the Third World: A comparative analysis* (pp. 71–84). Westport, CT: Greenwood.

Kahn, H., & Brown, W. (1975, December). A world turning point: And a better prospect for the future. *The Futurist,* pp. 284–287, 329–334.

Marchetti, C. (1986). Environmental problems and technological opportunities. *Technological Forecasting and Social Change, 30,* 1–4.

Maynard, M. (2006, January 24). Ford eliminating up to 30,000 jobs and 14 factories. *The New York Times,* pp. A1, C6.

Moberg, D. (1993, June 14). Late to the station. *In these times,* pp. 14–17. (Chicago Institute of Public Affairs)

Moore, C. (1995, January/February). The green revolution in the making. *Sierra,* pp. 50–52, 126–130.

Murshed, S. M. (1993). The North–South economic interaction and the environment. *Asia–Pacific Journal of Rural Development, 3,* 41–53.

Out of denial. (2002, June 5). *Louisville Courier–Journal,* p. A6.

Pimentel, D., Tort, M., D'Anna, L., Krawic, A., Berger, J., Rossman, J., Mugo, F., Doon, N., Shriberg, M., Howard, E., Lee, S., & Talbot, J. (1998). Ecology of increasing disease: Population growth and environmental degradation. *BioScience, 48,* 817–827.

Renner, M. (2000). Vehicle production increases. In L. R. Brown, M. Renner, & B. Halweil (Eds.), *Vital signs 2000* (pp. 86–87). New York: Norton.

Ritzer, G. (2005). *Enchanting a disenchanted world: Revolutionizing the means of consumption* (2nd ed.). Thousand Oaks, CA: Pine Forge.

Romm, J. (1991, July/August). Needed—A no-regrets energy policy. *Bulletin of the Atomic Scientists,* pp. 31–36.

Van der Straaten, J., & Ugelow, J. (1993, Winter). Environmental policy in The Netherlands: Change and effectiveness. *Dutch Crossing,* pp. 130–158.

Verrengia, J. B. (2002, July 22). Pollution blamed in killer drought. *Louisville Courier–Journal,* p. A2.

CHAPTER 14

Notes

1. It would seem that to have extended peace and unity in our society would require taking back much freedom and many rights and allowing only a set of common values. Because most of us do not want such a restrictive society, we accept the fact that built into our current society is—whether we like it or not—the potential for continual disagreement, dissatisfaction, opposition, and conflict. However, on further reflection, most of us would not want it any other way.

2. New technologies (e.g., computers), new ideologies (e.g., attempting to build a sustainable society and world [Brown, 1987]), and recent social movements (e.g., for African Americans, Native Americans, Hispanic Americans, women, gays, and the elderly) all have been major independent variables promoting social change in our society by producing new kinds of jobs, closer to equal opportunity for many groups of people, and new ways of how we can look at our world.

3. Karl Marx pointed this out in a number of his works. For example, see *Communist Manifesto* (Marx, 1947), *The Economic and Philosophic Manuscripts of 1844* (Marx, 1964), and *Capital* (Marx, 1967). For a more recent work on the creation of social structures by humans, see Berger and Luckmann's (1966) *The Social Construction of Reality*.

References

Berger, P. L. (1963). *Invitation to sociology: A humanistic perspective.* Garden City, NY: Anchor Books.

Berger, P. L., & Luckmann, T. (1966). *The social construction of reality.* Garden City, NY: Doubleday.

Blumer, H. (1971). Social problems as collective behavior. *Social Problems, 18,* 298–305.

Brown, L. R. (1987). *In building a sustainable society.* New York: Norton.

Durkheim, E. (1966). What is a social fact? In *The rules of sociological method* (pp. 1–13). New York: Free Press.

Ferrari, A. (1975). Social problems, collective behavior, and social policy: Propositions from the war on poverty. *Sociology and Social Research, 59,* 150–162.

Heilbroner, R. L. (1991). Three socio-economic capabilities for response. In *An inquiry into the human prospect* (pp. 77–121). New York: Norton.

Marx, K. (1947). *Communist manifesto.* Chicago: Charles H. Kerr.

Marx, K. (1964). *The economic and philosophic manuscripts of 1844.* New York: International Publishers.

Marx, K. (1967). *Capital.* New York: International Publishers.

Merton, R. K. (1967). Manifest and latent functions. In *On theoretical sociology: Five essays, old and new* (pp. 73–138). New York: Free Press.

Mills, C. W. (2000). *The sociological imagination.* Oxford, UK: Oxford University Press.

Index